PRAISE *for*

1,001 VOICES *on* CLIMATE CHANGE

"A hybrid of travel literature and oral history, Lockwood somehow shrinks the ungraspably vast problem of climate change down to a human scale, then patiently, carefully combines those individual voices into a planetary chorus. A monumental achievement."

—Robert Moor, bestselling author of *On Trails: An Exploration*

"'Tell me a story.' Is there a more fundamentally human sentence than that? Devi Lockwood circles the globe, seeking people's experiences with water and climate change, from cultural myths to rising seas' impacts on daily life to one woman's pain, tuned to the voices of the trees. Lockwood seeks and you, dear reader, shall find."

—Erica Gies, environmental journalist, science journalist, and author of the upcoming book *Water Always Wins: Going with the Flow to Thrive in an Age of Droughts, Floods, and Climate Change*

"In a world that needs more listening and more storytelling, Devi Lockwood covers the waterfront. This is an empathetic and beautiful book."

—Richard Louv, author of *The Nature Principle* and *Our Wild Calling*

"This dazzling and significant collection captures the voices of people around the world, from Tuvalu to Thailand, from Australia to Kazakhstan, who are experiencing firsthand the life-altering effects of climate change. Lockwood's approach to recounting their stories is compassionate and impassioned, focused as much on the tiny details of life as the larger planetary changes afoot in her interviewees' own backyards. *1,001 Voices on Climate Change* is beautiful and necessary reading."

—Amy Brady, executive director of *Orion*

"As the fight against climate change accelerates, Devi Lockwood reminds us why. *1,001 Voices on Climate Change* records vivid stories from those already living through the climate crisis. Lockwood takes us to every corner of the world to remind us to stop and listen. It is a compelling snapshot of this moment."

—Samantha Montano, PhD, author of *Disasterology: Dispatches from the Frontlines of the Climate Crisis* and assistant professor of emergency management at Massachusetts Maritime Academy

"In this book, Devi illuminates the human stories the world so desperately needs. Devi's gift is in meeting people as they are and pulling out the essence of their stories in such a way that speaks louder than words. It is not with spreadsheets, graphs, and technology that we will overcome the challenges of climate change, but with a transformation of our culture through story."

—Alina Siegfried, author, narrative specialist, spoken word artist, and systems change advocate

"A great storyteller needs first to be a great listener, and with each pedal of her bike—up and down previously unknown paths—Devi Lockwood hears from those living climate change and related water woes literally on the front lines. Her skills at storytelling are matched by her mastery of listening. The results are riveting."

—Bud Ward, editor, *Yale Climate Connections*

EVERYDAY STORIES *of* FLOOD, FIRE, DROUGHT, *and* DISPLACEMENT *from* AROUND *the* WORLD

1,001 VOICES *on* CLIMATE CHANGE

DEVI LOCKWOOD

Simon Element

NEW YORK LONDON TORONTO SYDNEY NEW DELHI

SIMON
ELEMENT

An Imprint of Simon & Schuster, Inc.
1230 Avenue of the Americas
New York, NY 10020

First Simon Element trade paperback edition June 2022

Interior design by Laura Levatino

10 9 8 7 6 5 4 3 2 1

Library of Congress Cataloging-in-Publication Data

Names: Lockwood, Devi, author.
Title: 1,001 voices on climate change : everyday stories of flood, fire, drought and displacement from around the world / Devi Lockwood.
Other titles: One thousand and one voices on climate change
Description: First Tiller Press hardcover edition. | New York : Tiller Press, 2021. | Includes bibliographical references. | Summary: "A journalist travels the world to collect personal stories about how flood, fire, drought, and rising seas are changing communities"—Provided by publisher.
Identifiers: LCCN 2021005829 (print) | LCCN 2021005830 (ebook) | ISBN 9781982146719 (hardcover) | ISBN 9781982146726 (ebook)
Subjects: LCSH: Human beings—Effect of climate on—Anecdotes. | Climatic Changes—Social aspects—Anecdotes. | Sea level—Social aspects—Anecdotes. | Water-supply—Effect of global warming on—Anecdotes. | Human ecology.
Classification: LCC GF71.L64 2021 (print) | LCC GF71 (ebook) | DDC 304.2/5—dc23
LC record available at https://lccn.loc.gov/2021005829
LC ebook record available at https://lccn.loc.gov/2021005830

ISBN 978-1-9821-4671-9
ISBN 978-1-9821-4673-3 (pbk)
ISBN 978-1-9821-4672-6 (ebook)

For all the storytellers

CONTENTS

INTRODUCTION

MARATHON

When I listen, the whole world widens.

I first felt the need to listen to strangers after the bombs. Two of them, pressure cookers filled with nails, exploded at the Boston Marathon finish line, killing three people, injuring 260.

Marathon Monday is a holiday in Boston; my professors had canceled their classes. On April 15, 2013, I rode my bicycle to Coolidge Corner to cheer on the runners. It was my first time watching a marathon up close. I pushed my bicycle as close as I could to the racecourse. Following the queue of people next to me, I dismounted and called out the names of athletes written on their shirts and arms.

"Go, Jason! Yes, Hannah! You got this, Raquel!"

They pumped their fists back in gratitude, smiling through pain. The raw human emotion, the joy of it, overwhelmed me. After an hour of cheering, I turned my bicycle north and rode over the I-90 overpass back to Harvard. My rowing coach, Michiel Bartman, was the first to alert me that something was wrong.

"You should come home if you haven't already," he texted. Something about explosions near mile 26.

I stared at my phone, shaken. How could someone hijack that joy? What could they have been thinking?

A few days later, Massachusetts governor Deval Patrick ordered all Boston residents to "shelter in place," putting the city on lockdown.[1] The streets of Cambridge were silent, empty. Occasionally a helicopter passed overhead, whipping up the air. A siren tore down a nearby street.

In my cooperative house of thirty-two students, everyone was anxious, jittery. Someone prepared a big pot of pasta, which we ate without looking at each other. After dinner we kept to our rooms, not wanting to see our fear reflected back at us.

CARDBOARD

When the lockdown lifted, all I wanted was to go outside: to walk and breathe and hear the sounds of other people. I craved looking at fellow human beings, face-to-face, without flinching. I needed to connect, to remind myself that not everyone is murderous.

The Wednesday after the marathon, biking home from campus, I found a bunch of blue and green balloons on the lawn of the Cambridge Public Library, left over from a Cambridge Science Festival event.

"Please take these," someone said, handing me the orange ribbon at the base of the balloons. I tied the bunch to the back of my bicycle and pedaled off.

In the kitchen, I poked around under the sink and disassembled a broccoli box. I found a paper grocery bag, cut it open, and taped the inside of the bag over the surface of the cardboard rectangle, covering up the vegetable company's logo to create a sign of my own. Next, I cut two holes in the top edge of the cardboard sign. I threaded a piece of polka-dot ribbon through the holes, tying knots and adjusting the length around my neck so that the sign fell over my chest. "Open call for stories," I wrote in Sharpie.

The next morning, I left the house early with the cardboard sign around my neck, the bunch of balloons in hand (to draw attention to myself, I reasoned), and an audio recorder in my pocket.

I remember the air was cool—a spring kind of cool that blushes with the promise of almost-warmth. As I walked with my sign, the streets I thought I knew transformed. Suddenly, every step was laden with possibilities. I could talk to anyone! Why not?

People stared at me, and my sign, and my balloons. Some paused long enough to make eye contact. Some approached me.

"Open call for stories?" they asked, reading my sign. "What does that mean?"

"Do you have a story to share?" I asked.

"What kind of story?" they said.

"Any kind," I replied. "And would it be okay if I make an audio recording?"

I met homeless Vietnam vets. A woman who lost everyone on her block to the earthquake in Haiti and was wearing a Lady Liberty costume, holding a sign advertising parking. I met a transit police officer who swore that his mother was dead for forty-eight hours and came back to life after he prayed, asking if he could just have one more coffee with her. A twentysomething on his way to a bar busted out a rap dedicated to me and my sign right there on the street—a friend backed him up with beatboxing. An inquisitive psychologist wanted to know what other people were telling me. One man told me his best friend was a clown. Another was worried about his friend who was making a choice between grad school in one city and a girlfriend in another. A woman who had written a poem that morning pulled it out of her purse to read it to me. A retired Spanish teacher swore to me that the Statue of Liberty was modeled after Marie Antoinette.

Back in class, I wrote a series of poems inspired by these stories. But the gift of it felt bigger. I wanted more. Once I started listening to strangers, I didn't want to stop.

I soon ditched the balloons because I needed my hands free.[2] I brought the sign with me everywhere. After a while, I felt naked without it.

KNEE

The year before, in July 2012, I tore my anterior cruciate ligament while playing pickup soccer in Argentina. I sprinted out of the goal fast in a Southern Hemisphere winter, turning left while my foot was still planted firmly to the right. One twist, one pop, and my body betrayed me. The scar from the reconstructive surgery sprawls across my right knee: a line no wider than a piece of spaghetti.

"At least your other leg works!" the players joked. I was on the turf, my knee unsteady. I took a taxi home, hopped up three flights of stairs, Skyped my best friend, and described the twist.

"You're going to be out for a while," she said. "This is going to be really hard."

I came home for surgery, woke up immobile. I was miserable. I wanted to be back in my body, but I couldn't. I could barely take care of myself. Just hobbling to the bathroom was a struggle. I needed something to get my mind off my current situation. I couldn't dance. I couldn't row. I couldn't walk. I needed a goal.

The first thing I could do to get my heart rate up in physical therapy was the hand bicycle: a machine where you sit and pedal with your hands. As I regained range of motion in my leg, I could pedal on a stationary bike. In recovery, riding my bicycle was the first thing I could do that allowed me to move my body fluidly through space. At first, I could ride my bicycle farther than I could hobble. And cycling was good for building up the muscles around my knee.

The lack of mobility that came along with recovery from this surgery in 2012 raised big questions. Who am I if I can't move? What do I most want to do when I heal? It was while healing from surgery that the idea for my bike trip was born. I thought to myself, *By this time next summer I'll be able to ride my bicycle a really long distance. I love rivers, so why not cycle down a river? And why not talk to people along the way?*

Over winter break, up late in my grandparents' basement in Canada, I started googling bike paths in the United States that follow rivers. I landed on the Mississippi River Trail. Why? Because it was there.

Bicycling is a kind of freedom. Balancing on two wheels, my body suspended in air, there is no windshield between me and the outside world. I move by the power of my own body. When it rains, I get wet. When I face a headwind, I push harder. When I ride in a tailwind, I float.

To save up for a touring bike, I spent six weeks in the summer of 2013 working at a camp for high schoolers in New Hampshire, co-teaching Arabic classes and saving up around $2,000. This, together with an Artist Development Fellowship from the Office for the Arts at Harvard University, was more than enough to get started. I set my sight on a forest-green bicycle: a Surly Disc Trucker with a steel frame and disc brakes, a durable rear-rack, and a set of Ortlieb panniers—waterproof.

MISSISSIPPI

In August 2013, I rode my bicycle eight hundred miles down the Mississippi River Trail from Memphis, Tennessee, to Venice, Louisiana, where the river meets the Gulf of Mexico. I traveled with an Olympus LS-14 digital voice recorder, the cardboard sign that read "open call for stories," and enough peanut butter and tortillas to make emergency meals for days. Along the way I collected stories from the people I met.

At times, I wore my cardboard sign around my neck and walked around small Southern towns with my voice recorder in hand. Other times, the fact that I arrived on a touring bike loaded with front and rear panniers was enough to pique interest from the people I met when I stopped to refill on water or food. I soon benefited from the kindness of Southern networks—people I had just met offered me their homes or called up their friends two towns away to put me up the next night.

Many people asked if I carried a gun. I didn't. In Vicksburg, Sandy Shugars insisted on buying me pepper spray from Walmart, "just in case." The scariest two things I encountered were a vicious nest of fire ants outside Vidalia, Louisiana, that chewed up my ankle when I paused one afternoon to find a place on the side of the road to pee, and the omnipresence of big dogs with a love for chasing bikes.

You might call this a solo bike trip, but I was never truly alone. People looked out for me wherever I went.

When I pulled off Highway 1 in Greenville, Mississippi, and into her driveway, Jessica Brent took one look at me and said "Girl, where are your lights?!"

The next day she came back from Walmart with an orange reflective vest, a pair of red and white flashing lights, and reflective stickers, which we stapled all over the vest.

People talked to me—all sorts of people. I listened. By the end of that first trip, I had compiled over fifty hours of stories that both fueled and inspired me. I didn't know exactly what I was doing at the time, but it felt right.

MOVEMENT

The Mississippi River is defined by its movement. Despite the best attempts of the Army Corps of Engineers, the Mississippi's banks are always jumping. John Ruskey, a river guide who makes his own dugout canoes, showed me a watercolor map of the river he had painted to illustrate this point—the layers of mud moving and twisting from bank to bank, leaving oxbow lakes full of fish in their wake.

I met fifty-seven-year-old Franny Connetti eighty miles south of New Orleans. When I stopped in front of her office to check the air in my tires, she invited me to get out of the afternoon sun. She had red spiked hair and a big smile. I felt welcome.

Franny shared her lunch with me. We bit into fried shrimp, the crispy flesh of it. In between bites she told me about 2012's Hurricane Isaac, which washed away her home and her neighborhood.

"We fight for the protection of our levees. We fight for our marsh every time we have a hurricane," she said. Despite that, she and her husband moved back to their plot of land, living in a mobile home, just a few months after the disaster. "I couldn't imagine living anywhere else," she said. They have since rebuilt their home on stilts.

"Do you think there will come a time when people can't live here anymore?" I asked.

"I think so. Not in my lifetime, but you'll probably see it."

To imagine the road I had been biking on underwater was chilling. Twenty miles ahead, I could see where the ocean lapped over the road at high tide. "Water on Road," an orange sign read.

Locals jokingly refer to the endpoint of State Highway 23 as "The End of the World."

Here was one front line of climate change, one story. What would it mean, I wondered, to put these stories in dialogue with stories from other parts of the world—from other front lines of climate change with localized impacts?

My goal, once I graduated, became to put stories of climate change in dialogue with each other, giving names and voices to those impacted. I wanted to humanize an issue often discussed in terms of numbers: millimeters of sea level rise or degrees of temperature change.

I applied to every source of funding I could find, and received many nos, but one yes, from Harvard's Gardner & Shaw postgraduate traveling fellowships.[3] With that, I was off.

Where to? My then-girlfriend advised me to go to Tuvalu. We were sitting in an alcove in a patch of sunlight, doing homework side by side. She was taking a politics of immigration course and tossed me an article she was reading about "climate refugees" in Tuvalu.

"You might like this," she said. I pulled up Google Maps and added a point.

What she knew about me, and what I have since come to know about myself, is that I am drawn to water. Water infuses my earliest memories and is a mental marker of the places I have lived: Lake Umpawaug, where the fishermen bushwhacked and flew. The Norwalk River, grit and cargo, ships and salt. The Squamscott, a tidal form of resistance training. The Charles River, where I rowed in college and came to know every sinewy turn—the way the water moves in different kinds of wind—the soft bend of a moonrise behind the Boston skyline as seen from the river basin, suspended between water and air.

Wisdom sits in places. So I went to some places, and I listened.

A man named Zorp in New Orleans, when I passed through in August 2013 on my bicycle, told me, "You know, poet, the thing is: you're going to find in your life, all the stuff you're doing, you're waiting, saving it up. Don't save it. Get out there."

I listened, Zorp. I'm doing it.

INSTINCT

One of the best pieces of travel advice came to me from Abby Sun, a filmmaker one year ahead of me at Harvard who spent the summer before her senior year jumping trains along the US-Canada border. She pulled me aside before I set out on the Mississippi River Trail.

"I have something to teach you," she said. "The Three Second Rule."

"What's that?" I asked.

"When you meet someone, decide in the first three seconds whether or not you trust them," she told me. "And go with your gut."

The times where things have gotten dicey have been when I haven't been able to exit a situation, or I haven't listened to myself. In following this practice on the road, I have honed my gut. I learned when to say no, and how to leave gracefully.

I recognize that my race and gender and class and ability (and even my height! I'm five feet, five-and-a-half inches, and hardly anyone considers me intimidating) afford me a privilege in these kinds of situations. Hardly anyone finds me threatening. People are quick to trust.

I have talked with men who have ridden their bicycles across entire continents and have never once been invited into someone's kitchen to chat. On the road, people are eager to help me out. It turns out that is great for storytelling, too. So many of the stories I recorded were in people's kitchens or living rooms.

And you can tell so much about someone by the way they organize their kitchen: Lynn and Graham Pearson in Whanganui, New Zealand, have walls full of spoons from all over the world—a collection that started with Lynn's first trip to the South Island as a teen; a recently divorced woman in Australia kept a guitar in the kitchen corner, the air a mix of cat hair and cigarette smoke; in outdoor kitchens in Tuvalu, I found the water carefully rationed and boiled in pots.

INVITATION

I traveled with a cardboard sign that read "Tell me a story about water" on one side and "Tell me a story about climate change" on the other. I traced out each letter with a black permanent marker and hoped that the message was clear. The sign, I'd like to think, is inviting. There's some kind of openness in its handwritten words, its quiet invitation to have a conversation, face-to-face. *Tell me a story.*

I have since gone through various versions of the sign, in seven languages (English, Thai, Lao, Chinese, Russian, Turkish, and Inuktitut). At first the sign was new and clean. Over time, it became full of life and wrinkles—an oil stain, a small rip. I loved watching it age. Occasionally, I taped up the edges.

I made a few replacements—once in Chengdu, China, at an office supply store near a university, another time in Balkhash, Kazakhstan, at an outdoor market. Cardboard is surprisingly easy to find the world over. And people have always been eager to give. I travel with a permanent marker.

Now, I feel comfortable introducing myself as a journalist. But at the start, this piece of cardboard was the easiest way to start a conversation. It gave me permission to ask questions.

As the journey continued, my equipment got better, too—I learned to record with a shotgun condenser microphone. (Yes, I get stopped at airport security lines constantly; no, it's not a weapon.) I also learned to insist on finding the quietest places possible, where a person's voice wouldn't be too echoey or overrun with background noise. Coffee shops, despite being full of delicious coffee, are the absolute worst places to do an audio interview. Milk frothing is a noisy process. So is bean grinding.

I shied away from taking videos or photos because I wanted to be able to make eye contact with people as they spoke to me. Something about sound is more intimate, I think, than image. The lack of a camera lens allows people to be more candid, too. When I'm in front of a camera, I can clam up. I think most other people do, too.

In a culture that values productivity over slowness, screen time over verbal storytelling sessions, holding a space for a story to be told, face-to-face, slowly, feels revolutionary. While water and climate change were the starting points of many conversations I had with storytellers, they didn't necessarily end there. The personal is political, and environmental. It's all intertwined.

DEEP LISTENING

Throughout the journey I developed a method of deep listening. This was hard-earned, and not instinctual. When I listen to early clips of my journey down the Mississippi, I cringe. I was too pushy. I spoke over storytellers, eager to insert myself. I listened only halfway—the other half of my mind on how I would respond or redirect the conversation next.

I am an auditory person. When I meet someone, the first thing I notice is the musicality of their voice—how they let the taste of a word linger on their tongue or send sentences flying into the ether. Breath. Intonation. Word choice. Sometimes my favorite thing to do is close my eyes and listen.

In Margaret Wheatley's 2001 article "Listening as Healing," written a few days after 9/11 for the magazine *Shambhala Sun* (now known as *Lion's Roar*), she wrote: "Great healing is available when we listen to each other. No matter what we have experienced in life, if we can tell our story to someone who listens, we find it easier to deal with our circumstances.

"Listening is such a simple act. It requires us to be present (and that takes practice), but we don't have to do anything else. We don't have to advise or coach or sound wise," she continued, "We just have to be willing to sit there and listen, and if we can do that, we create moments in which real healing is available."[4]

When I'm listening (microphone in hand, nodding along and not breaking eye contact), there have been so many times when people have said, "Thank you. Thank you for listening," or "It feels so good to share this story with you. I've never shared it with anyone else before."

In listening, I want to be sure that people feel respected and heard. I met a Belgian woman in the UK who told me the story of how her hometown's water supply was contaminated from a plant that put a waterproof treatment on fabric. A man from Afghanistan lost his brother because of water contamination. An American woman visited the Florida Keys as a teenager, only to return years later and find that the reef was dead, bleached. "Thank you for listening," she told me at the end, through tears.

I enjoy listening to people whose perspectives and takes on the world are different from my own. There's a difference between offering stories and opinion and offering something aggressive or vitriolic. The first comes from the desire to share and connect—the second is purely an attack. One carries the potential for change and exchange. The other does not.

Deep listening has to start from the basic premise that we are all equal, all worthy of being listened to, all human—that everyone has a story to share, that those stories matter, and that we can learn from each other, if only we are fully present.

Deep listening is listening without the intention to respond—listening with the whole of one's body: making eye contact, leaning forward, nodding along without interrupting. Deep listening is honoring, is bearing witness, is keeping one's ears and mind open without the distraction of ego or fear.

This kind of listening is urgently needed in the climate crisis. Because listening is the first stop on the way to solution building. If we're building solutions that don't take into account the voices of people who will be impacted, it's dangerous—and more importantly, ineffective.

SHAHRAZAD

Stories are the way we define ourselves in real time—a collection of narratives and ideas that transfers from my brain to yours, or vice versa. They are snaps, moments. No two storytelling events are ever the same. I could speak the same words I'm saying right now in Whanganui, New Zealand, or under a bridge in Abisko, Sweden, and the resonance would be different.

Some of the best stories are the oldest ones. The oldest surviving manuscript of *Alf Layla wa-Layla*, or *One Thousand and One Nights*, likely dates from the fifteenth century. But even before that, it was shared as an oral tradition for hundreds of years. The story flowed from Sanskrit to Persian to Arabic, adding new interlocking tales with each iteration. Even now, retellings of *One Thousand and One Nights* appear on television and films throughout the Middle East during Ramadan.

Picture a murderous king. Incensed that his wife had cheated on him, King Shahriyar ordered that she be executed, and then, as an act of vengeance, he proceeded to wed every woman in the kingdom, one by one, enjoy their company for a night, and then murder them in the morning.

But Shahrazad is a trickster. She sees a way out. When it's her turn to marry Shahriyar, she tells him a story that keeps him riveted all night but ends with a cliffhanger in the early hours of dawn.

The king, caught up in the suspense of her storytelling, decides to keep her alive for another night, and another night, until it's three years later, they have two children, and he has forgotten his desire to kill her altogether. This time, cumulatively, amounted to 1,001 nights.

The idea to collect 1,001 stories started as something of a joke. When I was in San Francisco in October 2014, passing through on my way to Fiji, I told Sophie Lee, a fellow cyclist: "What if I recorded a thousand and one stories?"

"You could do it," she said.

Hearing those words reflected back at me, I decided I would. This is the story of those stories.

VOICES

Why do I ask for stories about water and climate change?

They are interlocking issues. Climate change is difficult to visualize. Water is easy to talk about. Everyone needs access to safe drinking water in order to survive. And everyone has a story about water: witnessing a flood, or living through a drought, or even the experience of learning how to swim.

I believe in the power of stories well told. If only we can listen to human voices describing the impacts of climate change, I thought, it would motivate people to act.

Another answer to the question "Why water?" is that the substance is part of my primordial soup. When I was born, my mother worked as an aquatics director at a pool in Massachusetts. My earliest memories are of learning how to swim.

I remember how she taught me to float. She stood in the shallow end and held my shoulders and feet while we named shapes in the clouds: castle, bird's nest, mountain. After a few minutes she removed one hand, then the other.

When I closed my eyes, I could feel every inch of my body held up by the water. Water, for me, was a safe place, a place of togetherness.

I am also intrigued by water's dual capacity: for creation or destruction. Most people experience climate change through the medium of water—droughts, floods, intensifying storms, or loss of land along the shoreline, to name a few. According to the World Health Organization, 884 million people lack safe access to basic drinking water services worldwide. Climate change, urbanization, and population growth all pose challenges for water supply systems. By 2025, the World Health Organization estimates that half the world's population will be living in water-stressed areas.[5]

Safe and reliable water sources are a key component of public health. We all need water in order to survive. In 2010, the UN General Assembly recognized water and sanitation as a human right.[6] This international focus on water was further solidified when the UN adopted the Sustainable Development Goals in 2015.[7] These seventeen goals cover social and economic development on topics of poverty, hunger, health, education, gender equality, water and sanitation, climate action, and affordable clean energy. Many of these goals are interconnected. Advancing water sanitation and quality, for example, contributes to improving public health.

Due to poor infrastructure, millions of people die each year from diseases associated with inadequate water supply, sanitation, and hygiene.[8] Water scarcity, poor quality, and inadequate sanitation also negatively impact food security and educational opportunities. Drought hits some of the world's poorest countries, exacerbating hunger and malnutrition. Sustainable Development Goal 6 aims to ensure access to water and sanitation for all by 2030,[9] with a focus on increasing investment in managing fresh water and sanitation on a local level. Investment in water resources can contribute greatly to poverty reduction and elevate local economies.

Despite longstanding efforts to provide access to safe drinking water, sharp inequalities persist along lines of race and class. People living in low-income areas generally have less access to improved sources of drinking water than others of higher social and economic status. In some cases, this barrier can be a matter of life or death.

Critically, those who contributed least to the problem of global climate change bear disproportionate burdens of its impacts. These people are also the least equipped financially to deal with the fallout.

Human voices must be central to the way we discuss water and climate change. Part of my aim is to make water scarcity and climate change—a topic that can feel both scary and abstract—accessible.

But both domestically and globally, the conversation about climate change isn't including the right people, so I make a point of seeking out underrepresented voices on remote islands and at kitchen tables—to find the full texture and nuance that the conversation on climate change still lacks, even at this late stage.

1
TUVALU

FUN

The plane started its descent with only the ocean in sight. I pressed my nose to the oval window, trying to spot land.

"Is this your first time?" the woman next to me asked. I nodded, fidgeting with the clasp on my tray table.

"You're going to be uncomfortable," she said, placing her hands on top of her flower-printed *sulu*, a skirt-like piece of fabric that many Pacific Islanders wrap around their waist. She told me that she is Tuvaluan—that this is her home.

I closed my eyes as the nose of the plane angled closer to the waves. I could almost taste their lips, the orderly rows of arrival and salt. I bit down on the inside of my cheek. Approaching Tuvalu is an exercise in trust.

At the last possible moment, a strip of land appeared beneath us. The wheels rolled to a loud stop. We taxied past palm trees, a fence, many pens full of pigs, and concrete homes with tin roofs: gray-green rainwater collection tanks attached to each. I followed my row-mate off the plane, squinting in the sunshine, toward a one-room, open-air airport. The airport code: FUN. Three people on motorcycles idled, one foot balanced on the road, waiting for the plane to depart. A volleyball net billowed slightly in the wind. There is no fence between the runway and the country: seen from the air, the strip of runway is arguably the main geographic feature of Funafuti, a coral atoll and Tuvalu's capital, which sits 585 miles south of the equator.

A little more than ten thousand people live in Tuvalu. Generations ago, Polynesians navigated here by the stars, calling the sprinkles of land in the vast blue of the South Pacific home. With ten square miles of total area, less than five miles of roads, and only one hospital on the main island, Tuvalu is the fourth smallest country in the world. Disney World is four times larger in area. Tuvalu, formerly the Ellice Islands, became independent from the British Commonwealth in 1978; the flag still bears the Union Jack in the upper left-hand corner. The other three-quarters of Tuvalu's flag is an aqua blue, symbolic of the Pacific Ocean. The flag is dotted with nine stars, one for each island: Nanumea, Niutao, Nanumanga, Nui, Vaitupu, Fongafale, Nukulaelae, Niulakita, and Nukufetau. The vowels in this language taste as delicious in my mouth as the sunsets are bright.

Funafuti, the capital, houses about half of Tuvalu's total population. It has the feel of a small town—after a few days, people start recognizing a foreigner. Tuvalu receives some 150 visitors per year.

By some estimates, Tuvaluans will be forced, by water scarcity and rising sea levels, to migrate elsewhere in the next fifty years. This mass exodus is already happening. Large Tuvaluan outposts exist in Suva, Fiji; and Auckland, New Zealand.

I came to Tuvalu with a question: What does it mean for a whole nation to be on track to become uninhabitable in my lifetime? If there's no place like home—how does that definition of home change when home becomes unlivable for an entire country? How am I, as a white US American of part-British descent, complicit in Tuvalu's destruction?

My passport stamped, I wandered in the direction of the Filamona Moonlight Lodge, one of two places offering accommodation on the island. The other, the Vaiaku Lagi Hotel (now the Funafuti Lagoon Hotel), was far more expensive. I knew that Tuvalu is a cash-only economy, and operates using Australian banknotes, though there are octagonal fifty-cent pieces adorned with an octopus on one side and Queen Elizabeth's face on the reverse. There were no ATMs. I took out what I thought would be a suitable amount of Australian dollars before finally leaving Fiji for Tuvalu. I had planned to stay at Filamona, but the prices listed on the website were years out of date; I hadn't withdrawn enough Australian dollars to cover the price hike. This would only cover accommodation and leave me with nothing for food.

I pondered this predicament, sitting in the lodge's open-air dining room after dinner, watching a replay of a rugby tournament on Sky Pacific. The owner's daughter, Luma, a high school student, wrote some basic Tuvaluan words in my notebook: *wai* (water); *talofa* (hello); *fafetai* (thank you); *koe fano kifea* (where are you going?).

Three planes a week arrived, and one big ship each month from Suva. On my plane and also staying at Filamona was an eighty-one-year-old Japanese man who had retired twenty years earlier from Exxon in Tokyo. Tuvalu was his one-hundred-forty-second country. He spoke loudly and traveled with a camera, taking pictures that he would paint later in his apartment in Tokyo. Before I could object, he snapped a photo of me eating rice.

Funafuti is a skinny crescent of an island lined with hammocks, fishing nets, and kids. Before dusk, Luma took me around the whole island on the back of a blue moped. I sat on the back, holding her waist for stability as we rode out to the edges and then back again. At this time of evening, the runway was an all-out playground. People spiked volleyballs, played soccer and rugby, and jogged around the airstrip's perimeter. The airplane runway—a backbone of cement—was built in the 1940s by the US Navy during the Pacific theater of World War II. While dismounting from the moped, I burned the inside of my right calf on the exhaust pipe.

In the morning, I sprinted to the southern tip of the island to get away from construction and the whirr of the bandsaw; the lodge, I learned, was under renovation. The December day was already hot with the sun still low in the sky. The end of the road felt like a different world. I could see the sea on both sides and—at the apex—ocean everywhere. Puffy ice cream clouds floated on the horizon. I crunched over coral and rocks underfoot: loose and extraterrestrial. This could be the surface of the moon. Even at low tide, the water felt almost too close, the waves many hands, calling; an ever-present wind. This was one ending: flattened cans of V8 and beer scattered among the white pieces of coral, edges smoothed by the churning of the waves. I splashed a little sea water on the splotch of my exhaust-pipe burn.

Back at Filamona, I struck up a conversation with Makelita, who cleans rooms at the lodge. She paused sweeping and we whispered at length about her employer. She was paid AUD$25 a week.

Two days later, another plane landed. A piercing siren announced its arrival, warning both people and animals to clear the runway. I adopted the gestures of those around me, turning and looking to see who was coming and going. My Japanese neighbor waved goodbye as the incoming passengers emptied onto the tarmac. Later he disappeared in a speck, a contrail. After the siren diminished and it was safe to occupy the runway, I crossed to the other side to visit the meteorological station.

INTRUSION

Tauala Katea, acting chief meteorological officer, sat in his office surrounded by reams of paper and the pieces of various instruments that document the weather and the tides. He tilted one monitor to show me an image of a recent flood, when water bubbled up under the field by the runway. "This is what climate change looks like," he said. The first signs of change had emerged a decade and a half before.

"In 2000, Tuvaluans living in the outer islands noticed that their taro and *pulaka* crops were suffering," he told me. "The root crops seemed rotten and the size was getting smaller and smaller."

Taro and *pulaka*, two starchy staples of Tuvaluan cuisine, are grown in pits dug underground. This crop failure was the first indication that something was wrong.

Tauala and his team traveled to the outer islands to take samples of the soil. After a period of research, the culprit was found to be saltwater intrusion linked to sea level rise.

Since 1993, a tide gauge at the main wharf has taken regular measurements. The seas have been rising at four millimeters per year since the Australian government started monitoring in the early 1990s. While that might sound like a small amount, this change has dramatic impacts on Tuvaluans' access to drinking water. The highest point in the islands is only thirteen feet above sea level.

The last twenty years have marked a period of dramatic change in the Tuvaluan way of life. Thatched roofs and freshwater wells are a thing of the past. The wells have become salty, so they are repurposed as trash heaps. All the water for washing, cooking, and drinking comes from the rain. The freshwater lens underneath the island, a layer of groundwater that floats above denser seawater, has become both salty and contaminated. Each home has a water tank attached to a corrugated iron roof by a gutter. This rainwater is boiled for drinking and also used to wash clothes and dishes, and for bathing.

"We are no longer dependent on underground water," Tauala said, gesturing to the island around us. "The common areas where people depend on and fetch their fresh water out of it have become salty. So they just fill it up with rubbish."

Water shortages and intensified storms have become a normal part of life over the past twenty years. But this wasn't always the case. In the past, Tuvalu had a freshwater lens; Tuvaluans could dig a shallow well and have access to potable water.

Imported food is now commonplace, even in times when there isn't drought. During my month in Tuvalu (December 2014 to January 2015), I learned what climate change tastes like: imported rice, tinned corned beef, a handful of imported carrots and apples, the occasional local papaya, bananas, and many creative uses for custard powder. Cooking happens on gas stoves in outdoor kitchens or, when a pig is prepared for a special occasion, on coals underground.

"We mostly depend on imported foods," Tauala said. "It is hard." In recent years, he told me, king tides have become stronger, another cause for concern. A king tide occurs when the Earth, moon, and sun align in their orbit, combining gravitational forces to create the highest tide of the year.[1] During the king tide in Tuvalu, water bubbles out of the ground.

"People experience a swell of currents of the king tide. It's spreading more inland and further toward households," Tauala explained. "Here in the Meteorological Office, our surroundings, during king tide, will cover with seawater."

The Tuvalu Meteorological Service website once featured a picture, since removed, of seven Tuvaluans standing in ankle-deep water in front of the white box on stilts that houses surface weather monitoring instruments.

"The water has spread up and inundated most of the low-lying areas. Year by year you can see that there is an increase in a trend. It covers further inland. This is a new experience for us," Tauala said.

He also worries about storms that coincide with a king tide. Any cyclone, "when there is a direct hit," he continued, "it will be devastating."

Tauala told me that climate change "shifts the pattern of everything, even our rainy season." In the past, the wet and dry seasons were predictable, with above-average monthly rainfall from November to April, and less rain from May to October.[2] "Nowadays there seems to be a change in the rain pattern," he mused. "We seem to have frequent heavy rain. The total rainfall for each month during the dry season is above normal or below normal."

There is no normal anymore.

"We can try to adapt to climate change, all these changes," Tauala said, "or migrate."

For a low-lying nation like Tuvalu, the impacts of sea level rise can be devastating. Relying exclusively on crops planted in the ground is a thing of the past. Things that are imported run out. Rice? Stock up. Apples and carrots? Enjoy them for the two weeks while they last.

Farther down the runway from the meteorological office, the Taiwanese-funded Fatoaga Fiafia Garden plants salt-resistant seeds in raised beds; the produce sells out quickly on the two mornings a week they open for sale.[3] You had to know someone who works there in order for them to set aside a share. Family kinship is king.

STAYING PUT

From the meteorological office, I walked next door to the public works department, my cardboard sign flapping slightly in the wind, and made friends with Ila, age thirty. A new church was being built near the northern tip of the island, and her job was to oversee the installment of new rainwater collection chambers. On the roof, "there's lots of surface area," she told me, "so there will be lots of water."

I asked her if she had ever thought of leaving—of making home somewhere new.

"I go with the flow. If and when the time comes to leave Tuvalu, I will go with my family," she told me. "While it's a good idea to have a plan, for now I am staying put."

Ila thinks the elders who want to "live in Tuvalu and die in Tuvalu are foolish. Some hide behind the Bible, saying that a second flood will not come. And the only response to that is to smile and nod, right?" she said. "You can't really argue with a religion, or a religious conviction."

Ila was born in Suva and spent her early childhood there—her father was the head ambassador at the Tuvaluan embassy. She lives with her husband, a sailor who signs nine- or eleven-month contracts at a time. When he returns, he brings her back red or black clothes, her favorite. They have three kids: twin boys age five and a three-year-old girl. Ila had her first children at twenty-five. She dropped out of the University of the South Pacific in Fiji after two years because school wasn't for her, and has been working nonstop since age twenty. On slow afternoons in the office she plays offline games: solitaire or minesweeper. There were only three computers with office internet, and hers was not one of them.

When the workday was over, she invited me to her house on the other side of the airstrip so that I could use her Wi-Fi to let my family know that I was alive. Ila took a nap on a woven mat on the floor while I caught up with the other side of the world. Later that evening, as the sun faded to the west, her husband took the kids swimming in the lagoon. Ila and I went on a motorcycle ride to the northern tip of the island and beyond—over the causeway to the junkyard. The sunset was stunning. I wanted to throw myself into it.

Riding the moped in the dark, I could see Orion in the darkening blue: sideways and proud. In the center of town, the streetlights obscured it, but there was darkness just a few minutes away. With the sun down, it was finally cool enough to breathe deeply. The air churning in our faces was a welcome coda to a hot day. This time, while dismounting, I remembered to leave enough distance between my leg and the exhaust pipe to avoid a second burn. The previous one bubbled and peeled.

ADAPTATION

For Tuvaluans, adapting to climate change is, in some ways, the only option. When I visited in 2014, Soseala Tinilau was working as project coordinator for Tuvalu's National Adaptation Programme of Action (NAPA). He told me that one NAPA project involved planting trees along the coast; their roots can limit soil erosion. Another targeted an island experiencing seawater intrusion; NAPA helped them create an aboveground structure filled with compost and soil where the islanders could plant their root crops. Another Tuvaluan island, in an effort to bolster food security, received help to refurbish their inlet for fishing. A third was considering constructing coastal walls for protection from the tides.

"People in the outer islands, they really want to see solid solutions to their coastal problems like seawalls," Tinilau said. He pointed out that the highest point in Tuvalu is only thirteen feet above sea level. The country as a whole is ten square miles of land. "The low-lying atolls are pretty vulnerable to sea level rise, especially here in the capital," he said. "We are very much affected by climate change.

"You may have heard about the severe drought in 2011. Six months without rain. Everybody really suffered." In order to conserve water, people didn't shower. "They just bathed in the sea and that's it," he said, noting that such a lack in sanitation left Tuvaluans vulnerable to diseases. During that drought, there was a diarrheal outbreak in Funafuti.[4]

"We are very much affected by climate change," Tinilau repeated.

"The sea rises and comes from the coastal areas, but also bubbles up from the ground. When it does that, the salt intrudes in our water lens and affects our crops."

"We plant on saline soil, and you can't do that," he laughed, a toothy exhale. "All the plants will be weak or die."

"We are very much affected by climate change," he repeated for the third time, a refrain.

"Do you think there will come a time when the sea level has changed so much that people can't live here anymore?" I asked.

"Well, the scientists have been warning Tuvalu that maybe in fifty years' time, this whole land will be inundated. If you ask old people here, they have a different view because Tuvalu is a Christian country. They believe that God put us here for a purpose and He doesn't want to destroy our country. Things like that. That's old people thinking. But, you know, if you ask the youth, they will tell you that there might be one day that this land will not be habitable. It's pretty hard, but we just have to—we can't do much of anything about mitigation," he said. "So all we have is to adapt. Adaptation is the most pressing solution that we have so far to the changes in our climate."

"What are you most proud of in your work here?" I asked.

"Most proud of?"

"Yeah."

"Well, I'm a Tuvaluan," he said, adding that through his work, he could see that his fellow citizens engage in the conversation about climate change. "They accept shortcomings. They do not give up. That's what I take pride in."

For now, there isn't a national plan to leave. "Our leaders insist that we have to stay here and we need to adapt to the changes. Because if we move, then we've lost our sovereignty, our culture, our language. Moving to another country, it is pretty difficult to maintain all of this culture," he said.

"Fortunately, our leaders are dedicated. We were born here and we are staying here, so we have to make the most of protecting our islands from sea level rise," Tinilau said. "We do not wish to be relocated."

Instead, relocation is up to the individual. While most Tuvaluans wish to live where they were born, some people, especially of the younger generation, would rather leave. A migration program set up with the New Zealand government, called the Pacific Access Category Resident Visa, allows citizens of Fiji, Kiribati, Tonga, and Tuvalu who are between eighteen and forty-five years old to apply.[5]

Under this scheme, Tuvalu is given a yearly quota of seventy-five visas. If successful, married applicants can move with their spouse and kids to New Zealand. In order to enter the visa lottery, Tuvaluans have to have a full-time job lined up in New Zealand, be in good health (as proven by a chest X-ray and medical exam), and be able to read, write, and speak English.

Since the program began in 2001, Tinilau estimates that over a thousand Tuvaluans have moved to New Zealand in this way.[6] Many of them live in Auckland, though some are in Hamilton and Wellington. In some ways, this is an extension of the tradition of Polynesian migration. New Zealand's islands are larger; they can more easily absorb the impacts of a changing climate.

Among Tuvaluans, I learned, there's a generational divide in how people talk about leaving. Youth are taking the plunge; many apply to the Pacific Access Category Resident Visa until the odds shake in their favor. Others go to Fiji to study at the University of the South Pacific. Older generations see climate change as an act of God. They're staying put.

FAMILY

I met Alofanga while buying a scratch-off internet card to connect with home. She was working at the Telecom stand, selling plastic cards with data plans. I asked her if she knew of anyone who might have extra room for a stranger.

"Let me ask around at lunch," she said. "Come back in the afternoon?"

I walked out to a dock on the lagoon side of the island, where pieces of coral nestled like bones among the sand. Big white cumulus clouds lavished their rain elsewhere—I watched them drift close to the island, and then pass, their underbellies dark with water. Rains in Tuvalu come heavy in all-encompassing, single-cloud storms that are completely unlike the all-day, gray-sky rains I am used to in New England. Here, I learned, rain is a gift.

In the afternoon, I walked back up to the internet stand. Alofanga stood behind it, staring out into the street, empty except for a passing moped. "You can stay with us," she said. "Bring your stuff and we'll go."

I followed her out the back door and through a tangle of houses nestled close together, past outdoor kitchens under awnings and clothes strung out to dry beneath the roofs. Alofanga introduced me to her grandmother, her aunts, her cousins. A stand-up fan ruffled the edges of family photographs taped to the walls. I put my bags down. For a few weeks, this was home. Every morning I woke up at dawn to Alofanga's grandmother singing in Tuvaluan. One word I understood: *Jesus*.

"In Tuvalu, we have this saying: 'You cannot lift a cup with one finger.' In other words, we need our family to survive," Alofanga told me.

Of the eleven thousand people who live on Tuvalu's ten square miles of land, about six thousand live in Funafuti in densely populated homes. The islands are a culture of extended family; most nights, ten to twelve siblings, aunts, grandparents, and cousins crashed out on woven mats on the floor, competing for space closest to the fan, which pushed a spiral of hot air toward us as we slept. In Tuvalu, it's completely acceptable to fall asleep on the floor of whatever relative's house you happen to be visiting that night. Everyone shares rice and fish and fruit.

Alofanga introduced me to her cousin, Losite, who is two years younger than I am. I asked Losite if she could take me out to see the now-defunct wells.

"Sure." She shrugged. "But it's not much to see." We walked out behind a stand of houses shaded by coconut trees. A pile of stones stood at the center of a circular rock wall no taller than my shins. We peered in. At the bottom, some ten feet down, was a muddy puddle. I caught a glimpse of coconut husks and a potato chip packet.

"It's all garbage, sis," Losite said. "No water here."

While living with Losite and her family, I learned how to conserve water. Every drop from the rain was precious, and to be treated as such. Between rains there was always the question: Will there be enough water in the tank to get through the day? And if not, how can we cut back?

Alofanga taught me to turn off the tap while I lathered up my body and unss of curls. We snuck into the bathrooms at her workplace after —she had the keys, and no one else was using the water in those big tanks, she told me. The bigger the roof, the more water that funnels into the water tanks, and the Tuvalu Telecom building, a two-minute walk from her home, has one of the largest roofs in Funafuti (and blessedly, strong water pressure). The whole ordeal had an air of secrecy about it. We went after dusk and always entered through the back door. We never knew who might be watching. The island, a small community, has many eyes.

Under the Telecom's shower spigot, I relaxed. Cool water dripped down my back, washing away days of caked-on sweat.

The other solution to not showering is to dance. In Tuvalu, community members gathered after dark in open air halls to dance. Everyone in the audience brought a bottle of perfume (imported, of course). We sat on the perimeter of the hall, large and open to the breeze. People stood up to dance in groups of two or three. If someone liked the dancers, they walked up to them and doused them in perfume as they continued to move to the beat, drawing an *X* in the space around their bodies in motion. People cheered. The speakers were loud, stacked two or three high. The bass rumbled near maximum volume. The music carried us through.

LAUNDRY

I ask people for stories about water because we often don't know how to talk about climate change. The language of parts per million of carbon dioxide in the atmosphere is awkward and technical. I once heard Christiana Figueres, a key negotiator of the 2015 Paris Agreement, say that she wished she could paint every molecule of carbon dioxide in the sky a bright color. That way people would be able to see the impacts more clearly.

I didn't bring a paintbrush with me to Tuvalu, but I did bring an audio recorder. My mission: to listen to stories about water and climate change.

Why water? Water is a substance that we can see and feel. When water is not present in our lives during a drought, we experience its lack acutely. Water is rationed. In the most dire cases, our throats go dry. When water is present in overabundance in a flood or sea level rise, it becomes a force for destruction, washing away homes and businesses and lives. Many (but certainly not all) of the impacts of climate change are experienced through water.

Even under the shade of a coconut tree, it was hot. I felt sweat behind my knees and under my arms and at the nape of my neck—my body's attempt to maintain equilibrium. Each home, concrete and brick, has a rainwater tank affixed to the roof by a pipe. The sides of these tanks read: "Donated by the EU" or "Gift from the Commonwealth of Australia."

When the rain came—fast, thick, and percussive—I watched the most agile members of each household leap outside. They shook rainwater out of their eyes as they assured that the pipe was attached properly to their roof and tank. Water here is precious and threatened.

Once, during my month in Tuvalu, I asked Alofanga if I could wash my clothes.

"We have to wait for the rain," she said.

Later, I asked if I could help out after a meal by washing the dishes. She flatly refused. "White people don't know how to wash dishes without using all the water," she told me.

FOOTPRINT

In Tuvalu, 600 Mbps of internet in 2014 cost the equivalent of a week's wages (AUD$20). In Tuvalu, rain is seen as a blessing. In Tuvalu, I learned to take short, cold showers. Everyone got around by motorcycle; it was far too hot to walk for most of the day.

In my first week, Losite showed me around the island. We zoomed by houses and hammocks and fishing boats leaning against the trunks of coconut trees. A woman slept on a platform in the shade using a piece of coral as a pillow.

Funafuti is a semicircle. The protected side is the lagoon, a calm and sequined aqua that should be a color of its own in the crayon box. The unprotected side has hungry waves. At the skinniest ends of the coral atoll, I could see both sides of the ocean out of the corners of my eyes.

The carbon footprint of my flight from NYC to FUN emitted about one ton of carbon dioxide into the atmosphere. According to a World Bank data set compiled by the Carbon Dioxide Information Analysis Center in Oak Ridge, Tennessee, the carbon footprint of the average American is over fifteen tons per year.[7] The annual carbon footprint of the average Tuvaluan is just one ton.

Historically, the United States has emitted more carbon dioxide into the atmosphere than any other country in the world. But no one is asking me to move away from the place where my parents and grandparents live. Climate change in Tuvalu is an environmental justice issue—those most impacted by the problem are also those who have contributed to it the least. My wealth and my race and my passport insulate me from the worst effects.

In the evenings, Losite took me for joy rides on her uncle's blue moped. We zoomed past concrete homes with metal roofs and thatched meeting houses, waving at neighbors, until Funafuti ended at the wharf. Then we'd retrace our path, the churning breeze a welcome respite from a day of heat.

While we rode, Losite asked me questions.

"On your island, does it snow? On your island, do you have a boyfriend?"

Beyond the wharf is a dump, a curious collection of multicolored plastic detritus, hollowed out, skeletal ships, and some military equipment left over from World War II. At the opposite end—a rocky beach of rugged coral—water lapped on both sides of our peripheral vision. It takes only twenty minutes to circumnavigate the island from tip to tip in this way.

NUKUFETAU

Tuvalu is a nation composed of nine coral atolls, but only Funafuti has a functioning runway—the rest must be accessed by ship: the trusty, if rusty, *Navaga II*. Laden with supplies and family members, the *Navaga II* departs Funafuti's wharf for the outer islands once a month to deliver priests and family members and rice—all the essentials.

One of the outer islands, Niulakita, is so small that only a few dozen people live on it. A few days before Christmas, Losite and I boarded the overnight ferry to Nukufetau to visit her parents and siblings.

Nukufetau is a strip of sand even narrower than the capital. When it rains in the distance and the sun is at the right angle, a rainbow hangs in the sky, extending uninterrupted into the azure. Every direction away from Tuvalu is water.

On the *Navaga II*, which Losite nicknamed "the stinky boat," we ate coconuts and breakfast crackers, curled up on a woven mat on the deck along with the rest of the passengers, all Tuvaluan. The boat was delayed by eleven hours at the outset because we were waiting for a priest, and, of course, messengers of God have the power to delay departures.

We hit a storm that lit up the sky pink in the middle of the night. At some point, someone waved us inside a cabin with bunks. Losite periodically sprayed two types of perfume in the room. The upper and lower decks were filled with woven mats, unrolled, with people on top, fanning themselves until they fell asleep. I took a break from the room to get some air, standing on the deck of the *Navaga II*, watching the sea go by, the moon reflecting on the bumpy water.

At the wharf in Nukufetau we were met by Losite's mother, a primary school teacher; her father, who sells ice cream—which we ate for breakfast with corn flakes and sweetened milk—and her two younger sisters.

As we sat in the shade by her roosters, Losite's mother, Misikata Ielomi, told me more about the prolonged drought in 2011—six months without rain. She was concerned about the impact that climate change has on her community's health.

"We usually get water from rain," Misikata told me. "And when we fetched water from our tank in the drought, it was all empty. We depend on our underground well, and we also found the problem that the underground well water was salty. That was devastating," she said. "Some of us got sick and went to the hospital and got some medicines, and some even were transferred to the main hospital in the capital that is in Funafuti."

Misikata's pigs, another staple of Tuvaluan cuisine and life, suffered from a lack of food and water. Even the coconut trees weren't bearing fruit. It was "very hard to find anything to drink," she said, gesturing toward the trees that craned upward around us, leaning out over the water, their trunks like elongated straws.

During the drought, food rations and bottles of water arrived to each of Tuvalu's outer islands—donations from overseas. It wasn't always enough.

"We ask for future assistance," she said. "Whoever can assist us to deteriorate the problem, so that we can live a normal life and a happy life, and especially to our health. That's all I can say to you. Thank you."

SWIM

High tide after thick rain in Nukufetau. The water lapped close to my feet. I sat suspended over a trash heap—coconut husks and tin cans and empty chip bags—in a hammock, breathing in the horizon. If I squinted, I could see a rainbow. Tall clouds moved south, taking the raindrops with them.

Losite came up behind my shoulder. "Devi, we go bath in the sea!"

I wrestled my body from the view and followed Losite and her two sisters. We ran across the main road—a sand pathway—to the other side of the island. Soft sand squished under my toes. I could feel the spaces in the ground that were not long ago filled with rainwater. The whole trip took less than a minute.

Just before the beach, we passed the pens of many families' pigs, a mess of pink and brown bodies pressed close. For a moment, the smell of slop and pig poo permeated the air.

Then we reached the beach. Coconut fronds framed the red-orange curve of sunset. White sand kissed the mauve and azure of the water on this sheltered side of the atoll. We stepped in barefoot. The water was soft and warm.

Losite's middle sister, the daring one, started to swim across the bay to a small island. I followed. Losite laughed, revealing her missing front tooth. "I can't swim!"

I slowed down to stay with her, the vestiges of my mother's lifeguarding instincts taking over. I made sure that it was shallow enough for Losite to walk across the whole way. The tidal current pulled at our limbs, strong but not strong enough to knock us off balance.

When we were out in the middle of the water, I picked up a piece of algae and jokingly tossed it in Losite's direction.

Losite splashed her sister with a mouthful of salt water as she dodged the green, slimy projectile, prompting a full-on splashing and algae-hurling war. The rhythm of our laughter traveled across the water as we dodged the pieces of slime. I picked a fleck of algae off my cheek and exhaled salt.

We reached the opposite shore, but the bottom was sharp with crushed-up shells so we turned around. Back in the water, Losite's youngest sister pointed and said in a completely serious tone, "That's a big shark."

Losite and I freaked out and started hurling our bodies toward shore. The other two sisters collapsed in laughter. Cue more algae hurling.

The middle sister reached her hand to the bottom and picked up a tubular, brownish creature as long as my outstretched hand: a black sea cucumber. She offered it to me, and I touched it, recoiling and making a face at the slippery squish.

"This animal makes sand," she said, finding another and tossing it in my direction. I ducked, just in time for it not to hit my face. "We call it *loli*."

"*Loli*." I tasted the Tuvaluan vowels, the long *o* and the upswing of the *i*.

The dark rose. We continued to soak.

"Devi, on your island do you have a boyfriend?"

"In the States? No."

"Why not?"

"It's more fun to be alone."

"And on your island, Christmas is warm?"

"No," I laughed. "It's snowing."

"What would happen if it snowed for Christmas in Tuvalu?" one sister asked.

Losite thought this would be a brilliant idea. "We need snow! The sun here is too hot."

I expressed concern for the coconut trees. Could they handle it? Probably not.

The first star came out, then a second. I pointed and shouted every time a new one appeared.

"Two stars! Three! Four! Another, there!" The simple jewels.

"In Tuvaluan, we call the night sky *ceitu*."

I tried on the word for size. *"Ceitu. Cei-tu. Cei-tuuuuuuuuuuuuu."*

Losite and her sisters laughed and mimicked my exaggerated pronunciation. Drumbeats echoed across the water from the village.

"What's that?" I asked.

"The evening summon to prayer."

"That means it's almost Christmas!" I proclaimed. "Happy birthday, Jesus! Amen." I stuck out my hands like they were pegged on the crucifix and fell backward into the water. I dove down, did a handstand, and came up to cheers.

"Devi, how you do that?"

I explained putting hands on bottom, toes pointed in the air. The youngest sister tried and succeeded. The middle sister flicked her toes back and forth in the air like a mermaid tail. Losite tried but came back up for air, sputtering.

We gave up for home. Outside the one-room house we huddled around a spigot by the rain drum, passing around a bar of soap and a cup. We soaped our hair and used the suds to point it straight up in the air, giving ourselves Mohawks.

The middle sister said to me, "You know, Devi, you're a nice girl."

"Ha! And you're a very naughty girl," Losite retorted. She was met with a cupful of water in the face. We all laughed.

"Tomorrow we bath at the wharf?"

~

The next day, Losite and I stood side by side at the top of the platform, looking down at the ocean from above: a stretch of pristine blue lapping at the concrete pillars.

"You ready?" Losite asked.

"Is it deep enough?" I managed, curling and uncurling my toes.

"Yes! Yes! Very deep. Tide is high." Losite punctuated each word with a hand gesture and smile, our fallback if neither one of us understood the other. "We go?"

I swallowed my suspicion as another round of kids, some naked and some in shorts, pushed past us to jump in, chortling and hollering and making all manner of gleeful sounds as they descended to the water below. It can't be that bad, right? I poked my head over the edge to make sure they were all unharmed.

"Okay, we go."

We walked closer to the edge, looking down.

"One, two, three!"

My body was filled with blue. How can the sea be this warm in December? I had to remind myself that Christmas is always sunny and sticky hot near the equator, far from the white Christmases of my childhood in New England and Ontario. I let myself float on my back for a moment, taking in the milky clouds overhead. Tiny electric blue and yellow fish darted in and out of the currents surrounding our bodies.

We climbed the stairs, dripping and laughing. Losite and I jumped again and again and again until the harsh bite of the salt water found my nose and my fingers were wrinkled like tiny walnuts. The tide came in, almost filling the shore.

The Tuvaluan island of Nukufetau sits like a smile of a sandbar atop the ocean. There are no cars, only motorcycles. You can walk from one curved end of the island to another in twenty minutes, and that's if you pause to chat with a few neighbors along the way.

Without warning it started to rain in big, fat drops. The rain stopped as quickly as it came. I thought about the rainwater tank back at the house, topping up, drop by drop. I looked to the horizon, where a rainbow arcs from one point in the Pacific to another, so clear I could lick it.

A second arc appeared, lighter and fainter about the first. I shouted and pointed, flapping my arms like a baby bird who hadn't quite mastered the art of flying. "Rainbow! *Nua nua!*" No one else seemed to take notice. Rainbows must be a daily occurrence in Nukufetau, especially in the rainy season. Children continued to swim and splash each other, dunking their heads below the surface. One found a dead fish, yellow with stripes of white and black. The fish passed between the children's hands, salty. Its eyes fixed on no one in particular.

MERRY

Losite's family lives in a one-room concrete home that her grandfather built at a perfect angle so as to maximize the passage of the sea breeze. The floor is covered with woven mats that double as a sleeping surface and an eating surface.

Decorating for Christmas happened entirely in the two days before. We hung up a string of multicolored lights around the perimeter of the room and taped a gold and red sign that read "Merry Ch" to the wall. (I don't know what happened to the "ristmas.") Losite's dad cut a branch of a bush-like plant from the shore that, when stuck in an empty tin of biscuits filled with sand, served as a tree. We wound more lights around the bush.

Tuvaluans open their gifts on Christmas Eve. At dusk, Losite's mother led us in a long prayer followed by a longer feast of barbecued chicken and a heaping plate of the local root crop, *pulaka*, followed by another blessing.

After the last streaks of barbecue sauce had been licked from the plates, Losite and I, as guests, were in charge of handing out gifts. Each family member had one gift hanging on the tree. Earlier that afternoon, Losite's sister meticulously wrapped each item in leftover lined paper from a school notebook. She sealed the edges with packing tape, securing each with a handle that could be used to hang the present directly on one of the Christmas bush's branches. Each person's name was scrawled on the paper in permanent marker, all caps.

We tore at the paper in unison. Each of the kids opened a packet of corn chips and a lollipop, which they ate immediately. Losite's father and I both opened a tin of pineapple in sweet sauce. It felt decadent.

"Thank you," I managed, overcome with an emotion I couldn't quite name. I held on to the tin, the cool cylinder and ridges. "You didn't have to do this."

I added two gifts to the tree: a beaded necklace, maroon and gold, that I bought from a craftswoman in Suva, and a postcard from Manhattan Beach with some Australian cash tucked in next to a thank-you note saying how grateful I was to spend time with this family. I make a point of thanking every person who hosts me, whether in writing or in person. Losite's mom read the letter aloud and then handed it to her youngest daughter to read to everyone again.

After that: nap time on the floor—the real festivities didn't start until the early hours. From midnight until dawn, local kids gathered at the playground with pieces of tin that they used as percussion instruments, the soundtrack for an all-night dance party. Someone tooted a whistle. We shook it under the stars. I walked back home under an upside-down Orion and fell asleep, joyous.

WAI

Back in Funafuti, a voice fell out from the darkened bedroom—a voice like dried palm fronds brushing against each other in the wind. Loud. Tumbling. Urgent. The words were one big exhale, a gravelly sigh many years in the making.

Makelita, my host for the week, was out celebrating the first day of the year with her friends. New Year's Eve is a weeklong affair in Tuvalu. I peeked into the living room to find everyone else asleep, sprawled across the woven mats that line the floors.

I gulped and walked into the bedroom, guided by a single fluorescent bulb hanging in the adjacent kitchen and my desire to be of use. Makelita's mother croaked from the bed. I saw the outline of her body lying down, felt her eyes on me as I moved. I treaded lightly, as if that could help. No one was around to translate.

The woman started again with a stream of Tuvaluan. Rhythmically I heard the same two sentences over and over. I recognized one word: *wai*. Water.

On a table close to me sat a thermos of boiled water. I poured it into the thermos cup and walked closer. The woman belted out a few more sentences that I had no hope of understanding. She was ninety-five and I was twenty-two and the gulf of language between us was nearly uncrossable. Nearly, but not quite.

I handed her the cup. "*Wai.*"

Makelita's mother sat up slowly, so slowly. The pink, plastic sheets crinkled under her bodyweight, releasing a faint smell of urine. I could see the edge of the adult diaper under her *sulu.* She lifted the yellow cup to her thin lips, one hand quaking on the handle. I reached over to steady it and help her drink—three small sips in all. It was still dark, but my eyes started to adjust. Makelita's mother rested the cup on top of her bedside table next to a blue crocheted cloth. She pointed to the bare lightbulb on the ceiling and mimed with her hands turning on a switch. I walked the perimeter of the room until I found it. When the light came on she smiled and I smiled. Up close I could see the texture of her white hair, her skin like a piece of roti, pocked with splotches of brown.

Makelita's mother continued to speak. I continued to have no idea what she was saying. Finally I recognized two words. *Igoa* and *palagni*. The first is "name." The second is the catchall term for white foreigner.

I pointed to myself and said my name. "Devi."

"Dafi?"

"Devi." I pointed.

"Dave?"

"Devi."

"Deni?"

We continued like this for some time.

Finally I pointed to myself and said, "America." Makelita's mother nodded, as if this was the answer she had been waiting for.

"Window." She gestured to the latch that would shut the slats of glass.

"Window!" I hopped up to shut the three windows.

Makelita's mother smiled. I smiled. She motioned for me to leave the room and I did. I looked in later to find her huddled in a fetal position, eyes poring over two pieces of paper from a daily prayer booklet. She mumbled the prayers aloud, clutching them like a talisman.

Half an hour later I passed the doorway while brushing my teeth. Makelita's mother was asleep, snoring lightly, the prayers on her bedside table next to the glass of water. I darted in to flick off the light.

The next morning Makelita threaded the sewing machine at her dining room table, readying herself for a day of hemming dresses. Trying to sound casual, I asked: "What's your mother's name?"

"My name?"

"No, silly, I know your name! Your mom's name."

"Oh, my mum? Her name is Malaika."

"Malaika. Malaika."

ILL

Sometimes, the water story took place in the landscape of my own body. For three days I was knocked out with a particularly virulent form of giardia, likely from drinking a bit of contaminated water. I made six miserable trips to the toilet on the first day, read the first book of *The Hunger Games*, and slept without memory of when I woke up and when I dozed off again.

I was vaguely conscious of the episodes of *MasterChef Australia* that played on a loop in the next room. Not one of these ingredients that the chefs are working with, not one, could be found on this island, I thought, bitter and nostalgic for grocery stores. The thought of food turned my stomach, though, and I rolled off to sleep. Makelita offered me tinned meat for supper, which I politely refused.

The second day was Sunday, and I knew I ought to tell Makelita why I was sleeping so much and kept refusing her offers for meals. She returned from a post-church feast at her sister's house with a plate piled high with rice and chicken in soy sauce, topped with a generous drizzle of ketchup.

My burps turned sulfurous overnight—this was when I knew it wasn't just the stomach flu. I had to seek out help. Makelita set the food at the edge of the mattress. The smell turned my already-queasy stomach.

"Makelita," I started, taking a deep breath. "I'm sick."

I watched her face turn to concern.

"I don't want you to worry about me, but I've had some awful diarrhea."

"You go to the toilet a lot, yes?"

"Yes."

"You should try eating something."

At dinner I managed rice. Just rice. Three bites, maybe four. Makelita rode her scooter over to the hospital—the one site of medical care on the island—to get rehydration salts for me from the dispensary. I learned that health care and medications in Tuvalu are free.

That night I threw up three times in a row, all rice and water. I stepped over a frog and two cockroaches in the hallway to make it to the toilet in time. There was no one to hold back my hair, but fortunately I had the foresight to cut it short a few days before getting sick.

A cousin once told me, "Devi, you know you're an adult when you clean up your own puke." The single bulb above the toilet wasn't functioning, nor was the flusher, and there wasn't a bucket of water nearby to dump down. I was completely beyond hauling a bucket in from the navy-green rainwater container out back; I barely had enough energy to make it back to the mattress in the hallway next to the kitchen where I collapsed. I took comfort in the fact that I have good aim and hoped that my hosts wouldn't be angry. Before I dozed off, I made a resolution to go to the doctor in the morning.

One of Makelita's nephews laughed out loud to episodes of *Mr. Bean* playing in the next room. He must be nocturnal.

On Monday morning, Makelita was scheduled for work cleaning rooms at the lodge. I woke up just before she left at 7:30, light-headed from dehydration.

"Makelita, today I'm going to the hospital, okay?"

"Okay," she nodded, halfway out the door. "Can you close the back door for me?"

"Of course."

It took me a full hour and a half to leave the house, though, because I kept surrendering to mininaps. The temperature climbed and I sweated even sitting in front of the fan. A line of Audre Lorde floated into the haze: "Caring for myself is not self-indulgence, it is self-preservation and that is an act of political warfare."

Move.

I instructed myself to leave in small steps that seemed more manageable than the twenty-minute walk to the hospital. Devi, sit up. Put two feet on the floor. Good. Put your backpack on. Take it off. Lighten the load a bit. No, you really don't need a laptop or charging cord at the hospital. Good. Sling it back on your shoulders now. Walk the ten steps to the front door. Put on your right shoe. Put on your left shoe. Take a sip of salty water. Take another sip. Now walk.

I took the journey in small chunks, moving slowly. While I had never been to the hospital before, I knew that there were only two main roads on the island. It would be difficult to get lost. I stopped by a baker to double-check my directions and bought a bun that turned out to be a frosted piece of white bread. I surprised myself by managing to keep it down.

When the woman at the shop saw me sitting at the bench under the awning a good ten minutes after she sold me the bread, she asked, "Did you like it?"

"The bun? Oh, yes. It's quite a feat, too, because I'm sick." I licked the icing from my fingers. "Do you make them?"

"Oh, no. My husband does. He bakes and I sell."

"He must wake up very early in the morning. The bun was so fresh."

"Yes. Always early. Bread in the dark," she said.

I thanked her for the fifty-cent bun and trudged onward toward the hospital. The waiting room was full when I arrived.

"ID card?" the woman at the front desk asked.

"Yes, I'm here to see the doctor."

"ID card?"

I pulled out my passport and handed it over, grateful that I remembered to bring it, given my current physical and mental faculties.

"D-e-v-i-K.-L-o-c-k-w-o-o-d" she typed with her middle finger. "You take a seat now."

"How long will it be? An hour? Two?"

She smiled, pitiful. "Two or more. Please sit."

I managed to find a seat at the edge of a bench next to a breastfeeding mother and child and two teens playing cards. The room was clamorous with children. The walls were green. A kid two benches over spit out a red lollipop and wiped it off on his shirt. The adults in the waiting room seemed absent or occupied with sicknesses of their own.

Eventually the exertion of sitting upright was too much for me and I had to lie down. I found a spot next to the windows in the back of the room and curled up, using my backpack as a pillow. The last thing I remembered before falling asleep is a Tuvaluan woman pointing me out to her neighbor and saying: "*Tapa* (Wow)."

I opened my eyes, disoriented. Green walls. The sun was higher and the room was hotter than I remembered. To my right a group of Tuvaluan kids sat under a crumbling stairwell, tossing bits of tile at each other. One hit my leg with a clattery *thwack*. I half fell asleep again, only to be woken up by another flying piece of tile.

Mercifully the door to consultation room number five swung open and someone with a clipboard called my name: "Dave Lockwood." Finally.

I wasn't sure where to sit. A Tuvaluan woman with a topknot and colorful scrubs stared at a computer screen with my name in the top right corner.

"What's your problem?" she asked, still looking at the screen.

"Diarrhea."

She looked me over from head to toe and laughed, a full-bellied, hearty chortle. I blinked twice, refusing to believe that a health-care professional was laughing at me.

"Ha—okay. Diarrhea." She typed the word painfully slowly, using only her right pointer finger. "And how many times do you, you know . . ." she trailed off.

"Anywhere between four and eight times a day. Oh, and last night I threw up three times."

"You puked?" her face contorted into pure disgust. I had to remind myself that I'm not paying for this.

I took a deep breath. "I need Flagyl. I used it once before in Tanzania and it worked for giardia. I have giardia."

"What is Flagyl?"

"Metronidazole. Do you have that?"

The nurse-doctor spent a few more moments typing before looking up at me. "Yes. We have. I give you one dose? Five days enough?"

"Okay."

She sent the document to the printer and handed me the prescription, signing it with a loopy hand.

"Where do I take this?"

"Dispensary. That way," she pointed. "Next."

I ambled in the direction her fingers pointed toward, finding a door marked Pharmacy. It was locked. I put my prescription in a pile along with thirty or so others at the window next to the door. An hour later a harried-looking man unlocked the door and scooped up the pile of prescriptions, disappearing into a back room. I wasn't the only one waiting, but I was beyond striking up a conversation. When the packet of pills with my name on it arrived at the front window, it was late afternoon and I felt that I had witnessed a small miracle.

Back in Makelita's house, I was startled awake in the predawn hours by, of all things, a rat crawling over my leg. The lights were never fully out in the hallway where I slept, so I could see its furry body clearly, the beady eyes staring up at my own.

I panicked—"Jesus!"—shoved it away, and brought my pillow into the room where Malaika sleeps. I curled up on the floor next to Makelita and her two grandsons.

How do you say, "Tuvalu, you kicked my ass," in Tuvaluan?

Tapa.

MOTHERHOOD

For many Tuvaluan women, scarcity can feel like fear. Angelina, a Tuvaluan mother of three from Funafuti, was honest with me about the impacts of a drought: "If there's no water, we worry a lot," she said.

In 2011, Angelina's middle daughter, Siulai, was only a few months old. Angelina told me that she, her husband, and oldest daughter could swim in the sea to wash themselves and their clothes. "We only saved water to drink and cook," she said.

But what about her baby? The newborn's skin was too delicate. If she went into the salt water, she would get a horrible rash. How can a mother decide between having water to drink and water to bathe her child?

ECONOMY

I learned from an Australian expat working in the Tuvalu Government Building selling fishing licenses that Tuvalu's economy is supported in part by leasing the .tv domain name. I watched many boat lights twinkling offshore, knowing that the nutrition of the fish caught on board was going elsewhere. There were outboard rigger canoes leaning against the trunks of coconut trees outside some homes on the lagoon. But in my month in Tuvalu, I never saw a local fisherman out at sea.

Pigs are a common source of protein. Every night at dusk, someone in the family takes a motorcycle or walks across the runway balancing two buckets full of the slop of the day. Those buckets made repeat appearances in times of drought—a means of carrying water from the desalination plant.

MOVEMENT

Tuvalu's one dance club is located in a converted chicken farm. Angelina brought me here, insisting that I had to celebrate my last weekend in Tuvalu. Four-foot tall speakers blared the great classics at max volume—*"Move, bitch, get out the way"* and *"Turn down for what?"*—plus an improbable mix of Latin pop songs that Tuvaluans sing along to.

I downed a beer, made peace with necessary hearing loss, and dove onto an empty floor to dance with Angelina. In less than twenty seconds we had a crowd of men vying for our attentions. They grabbed my elbow, my wrist, and held fast.

"Stop! Let go!"

I struggled against their grip, but it was futile.

One guy, Angelina's cousin, seemed kind enough. Almost. His frame was an inch shorter than my own and he didn't grab at any part of my body. I used him as a shield against the more aggressive men.

We danced for three songs, after which Angelina left me in his care while she jetted home on her motorcycle to check on her one-and-a-half-year-old daughter. When she was out of sight, Angelina's cousin grabbed my wrist (there, the concealed force) and led me out to the airstrip. I focused on naming the constellations above the tarmac, making up new ones when my knowledge ran flat. A kite unrolled spools of footprints onto the sky. Rock candy. Orion. My whole body tensed.

I was hyperalert of shadows on the runway, a pack of wild dogs, couples sitting with their legs intertwined. The cousin asked where I was from and if I was married.

I told him I was not interested in men. He pressed this point.

"Why not? Why don't you come closer?"

He scooched his bum toward mine on the gravel. I scooched away.

"I'm just not. Please respect that." I considered dropping the L-word on him but didn't.

He talked about himself to fill the space. He works at the Public Works Department, installing water catchment systems on the island's north end. He is nineteen. He has two sisters and a brother.

Keep him talking.

He has never heard of Boston. Everything he knows about America comes from Hollywood. He likes action films best.

"Does everyone have a gun in America?" he asked.

Angelina came back in time for the three of us to ogle the rising moon in the east, waxing gibbous, orange and low on the horizon, open like an eye, staring back.

CONTACT

For a few years, on Facebook Messenger when our time zones aligned and her green circle was on, Angelina would ask me if I had settled down yet.

"Sis," she wrote in 2017, "when are you planning on settling down? If u happen to have a boyfriend make sure to tell me. Love to hear the news."

I didn't have the heart to come out to her and intuited as much. Tuvalu, I later learned, has a fourteen-year penalty for homosexuality. Sometimes silence, when traveling alone as a woman, is the easiest thing.

GIFT

When someone leaves Tuvalu, departure gifts are common, because when people leave the island, it's uncertain when they will return. Angelina gave me a star-shaped necklace of shells that she had beaded herself. Losite's family gifted me a red-and-yellow oval fan—woven with pink, green, and orange plastic feathers around the perimeter—that was supposed to have my name on it.

On one side it read "Tuvalu"—on the other, a misspelling of my name: "Die."

I was about to board a seventy-two-passenger plane back to Fiji. It made me laugh. Losite's family was extremely apologetic, but the fan strikes me as a metaphor, too—I still have it, tucked in an envelope beside my desk. Cultural survival, climate change, death and life are intertwined.

Can an island culture continue to exist when fresh water, the very substance that makes life on an island possible, is under threat? What will it mean for Tuvaluan culture to morph, to migrate—to exist in a new place? What aspects of Tuvaluan culture will die as a result of our changing climate and the water scarcity that it brings? What will survive?

In Tuvalu, Alofanga told me, it's customary to bury family members in the sand in the front yard, because there is nowhere else. Ancestors don't migrate; they stay. It's difficult to leave their bodies behind.

I boarded the plane. A siren alerted the stray dogs and midday motorcycle riders that departure was near. The engine kicked in. We taxied and climbed.

I pressed my nose to the window, not afraid for myself, but for this circle of islands and the Tuvaluans who call them home. Funafuti became no larger than my fingernail, a dot in retreat.

Losite, in the years since I visited her, moved to Fiji, two hours by plane away from home, to study at the University of the South Pacific. She gave birth to a daughter; she looks more like her mother than when I saw her last. Losite is studying to become a primary school teacher, just like her mother. After finishing her degree, she will return to Tuvalu to teach. Her father, the ice cream seller, passed away. Her mother now lives in Funafuti.

Angelina and her husband, after two years in Fiji, are back in Tuvalu and waiting for the visa lottery to go to New Zealand. Her daughters are eleven, ten, and seven years old, and she recently gave birth to a baby boy.

There is a sizable Tuvaluan community in Auckland, but Angelina knows that her older relatives will not join her there. She doesn't reckon that many people in her extended family will apply to emigrate. Part of that is because it's difficult to leave family members, both living and deceased, behind. Being close to ancestors is as important for survival as having access to fresh water to drink.

"My brothers and sisters are all in Funafuti. Auntie too. And Uncle. I don't think they are moving, because it's very hard to move," she told me. "My mother and father who have passed away are resting there. It's very hard to leave them."

With displacement, some part of Tuvaluan culture will wither.

One day, archaeologists might excavate the layers of trash archived at the bottom of a salty well in Funafuti, deposited like layers of sediment, and calculate the moment when the sea change began.

2

FIJI

KAVA

The rain turned on and off like a light switch, pinging from fierce to eerily quiet. One exposed bulb looked down on the ten faces huddled on rickety wooden benches. The wood creaked and sighed when someone shifted their weight. We were on the porch drinking murky cups of grog from halved coconut shells. It was a Friday night in Lami Town.

Ateli, a man in his fifties, scooted closer on the wooden bench to tell me a story.

"Climate change, what do you call it, rising tide?" he began, pausing to slip a coconut shell full of grog between his lips. "You all think it's because of pollution—problems with the environment."

I nodded.

"That is not a Fijian problem. In Los Angeles, they have pollution so bad you can't even see the city from the water. But not here in Fiji. Our problem is with cement."

"Cement?" I asked.

"Cement," he continued. "You look around these village houses and they are all made of cement. It didn't used to be that way. People used to make their houses from wood. Bamboo. Now we use cement. This technology is no good. Where do we get the cement from? The sand. The villagers take sand from the shore. They take and take," he said, wrinkling his brow. "And it is the little bugs in the sea who are responsible for making the sand, you see, the little fish. The sand is what they leave behind after they eat."

Here he accelerated—one continuous exhale, punctuated by his outstretched arms, gesturing to the width of the shore. "But the rate at which we take out the sand to build cement and the rate that the little fish leave it behind, there is no comparison."

Someone passed Ateli another coconut shell full of grog. He drank it in one gulp, then passed the bowl to his left. A bead of sweat rolled down the back of my leg.

"So they take the sand and the water comes in, closer to the homes they just built," Ateli said.

It was my turn to drink. Ateli clapped three times as I gulped down the brownish liquid.

"People say that the islands are disappearing because of rising tides. But it's not about the environment up there in the air. It's how we use it down here on our land."

He wiped his lips on the back of his sleeve.

"The sea will come and take back what we have taken."

RUGBY

While I was on the road, a chain network of hospitality kept me alive—a string of generosity that I am eager to pay forward. You only need one act of kindness to set the magic in motion. You only need one connection to get started.

My stay in Fiji started with one friend, Margot Leger, who had traveled on a similar grant from Harvard the year before to study methods of traditional boat building in Fiji and Zanzibar. Margot introduced me to Aaron March, a photographer, who collected me, jet-lagged, from the airport in Nadi. I deplaned, disoriented. Waiting for Aaron, I put on my cardboard sign: "tell me a story about climate change." A man who worked at the airport approached me with a greeting.

"*Bula*. My name is Rudi," he said in a deep, resonant voice. We found a table above baggage claim to sit across from each other. Rudi told me that in his lifetime, he had seen the water level rising at Wailoaloa Beach, just a ten-minute drive from where we sat. He said yes to turning on the mic.

"I remember when I was younger, the shore was a bit wider and we could even play rugby when the tide was low," he said. "But now there's only space to sit and play in the sand and get into the water. We can't even play rugby on the shore anymore."

The lack of consistent precipitation had also been noticeable. "We haven't had rain for quite some time now," Rudi added. "But in Fiji, we are survivors. We survived four coups and we are always okay, because we believe in God, and we are so grateful. He will take care of our tomorrow."

For Rudi, survival and godliness are intertwined. He wanted to see climate action from his country, though he didn't specify what form that might take.

"We should do something," he said, "and help the whole world in whatever actions they're taking as well."

GUNPOWDER

Nadi had a very particular smell: toasted spices, warm, damp earth, and salt. The scent was there as soon as I stepped off the plane and onto the jet bridge, and it only intensified throughout the day, baking to a steamy crisp. For the first few days I felt untethered, drifting. I needed to figure out things like phone and internet and where the heck I was going to go. I was afraid of making mistakes.

"Never touch another person's hair," Aaron told me over a plate of dal and fried chicken. "The head is a sacred space."

My first full day in Fiji was Diwali, the festival of lights. By midafternoon, our downstairs neighbors delivered fried peas, sesame sticks, oblong doughnut holes, and balls of rice with cardamom and raisins. Aaron and I split each treat with a knife. The sun arched toward the west, piece by piece, lengthening to blue.

The fireworks started before dark. I heard what I couldn't see: the pop, hiss, and crackle of light almost touching the clouds. Aaron and I stepped out onto the porch of his apartment in Martintar, a neighborhood in Suva, camera in hand. He showed me how to decrease the shutter speed and f-stop to capture the essence of each explosion. A rooster crowed, cacophonous. Tall, proud booms echoed off the hills and buildings as people set off fireworks in different parts of the city. Three planes arrived, churning the air around them. Laughter wafted up from the first floor. Glasses clinked. A radio played "Bang Bang" by Jessie J, Ariana Grande, and Nicki Minaj.

"Flight attendants live down there," Aaron said, sensing the direction of my listening. "Nadi is a passing-through town," he added. "No one stays here for long."

Aaron boiled water from the kettle and drained it into an empty handle of Absolut vodka. "The water in Suva is safe to drink," he said, "but we must boil it first. If you drank from the tap, it would be okay. But boiled is best."

Between my airplane-dry throat and the general heat of the city, we refilled the 1.75 liter handle several times throughout the day.

I watched through the window as a woman in a sari on a second-floor apartment lit a *diya*, or oil lamp, and placed it at the edge of the porch. A man looked on from an outdoor couch under the balcony.

"You smell that?" Aaron took a deep breath.

"Gunpowder," I replied. The whole of the neighborhood was thick with firework residue. A haze of it sat over the rooftops.

I fell asleep facedown in his guest room. The fireworks followed me in dreams.

LEMON LEAF TEA

From Nadi I boarded a bus to Suva, the capital, where I stayed with Esther Sue, a banker a few years older than me—another connection from Margot. Inside Esther's family's home, I found a clutch of fierce women: Popo, an outspoken nonagenarian; Esther's Auntie Annie; and her sister. They taught me how to make hot sauce and lemon leaf tea. On Halloween, for a cultural mash-up, Esther and I carved pumpkins together on the porch, our faces illuminated by the candles we plopped in their hollow insides.

Nourished and rested, I started my story walks radiating outward, wearing the cardboard sign. I didn't know exactly where I was going to go, but I figured it out by following a combination of instinct and luck. Soon, those walks radiated to boat trips to the outer islands. I didn't bring my bicycle with me to Fiji or Tuvalu, on account of all the water—it was easier to get around without it. I started each day not knowing where I would end up; the general directive was to listen. It was my favorite kind of wandering: chasing the flow state.

HUMAN GEOGRAPHY

One day, the flow led me to the University of the South Pacific. Walking down Grantham Road, I was tugged into a several-blocks-long conversation with two guys on their way to class.

"Do you want to see campus?" one asked, readjusting the weight of his backpack on his shoulder. I shrugged. Why not?

We passed the school's outdoor café and the library with its slatted windows, walked over a trickle of a concrete river that they told me swells in the rain, and uphill to a clearing. Two wide-canopied trees offered their shade over a smattering of stone picnic benches.

My walking buddies ducked into a lecture. I stayed outside, pausing to take in the different social groups, the pop music they played through their phones, the hellos they called out to people on the path. Before long, I was spotted.

"Hey, you with the sign!" one girl called, waving me over to her tableful of friends. The group stopped talking when I got close enough to be in earshot.

"Hello. Welcome to the best table in Fiji. What's this all about?"

I explained that I had just graduated from college, and about my journey to collect stories from people about water and climate change. I traded a stick of gum for a few potato chips. Two students took on the task of telling me about Fiji's volcanic past and changing coastline. They interviewed each other—I didn't need to intervene.

"I'm taking human geography," one student said.

"Humans are part of nature!" another added, taking on the role of interviewer. "What is climate change?"

"Climate change is not that evident in Fiji, but it's happening all over the world. It's a really big issue."

"Does Fiji suffer from climate change?"

"Yes. It's becoming really hot," the student said, explaining that sea level rise, while a problem on the outer islands, isn't as much of a big deal on the 'mainland' (the largest island) "because it's really high."

"Our seasons have changed in a really big way. Our summer is actually quite cold now. Before summer used to be hot."

"And when it's hot, it's just really hot."

"Some of the international students have said that they came to Fiji for sunshine and sandy beaches and now it's raining. They didn't come to Fiji for this. It's quite a change."

"This is really really really cold for us," one student added. "It's not normal."

"How about the islands? How about in Kadavu? Have you experienced the rise of the sea level?" the student asked.

"Yeah. It's evident. You know how you have the coconut trees next to the beaches? It's all inundated with sand now. It's no longer like this, how you have the coconut trees and the beach and then the sea. Now it's just the coconut trees."

"You mean there's a lot of sand, but there's coconut trees in just a small portion?"

"Yeah."

"You can say that's probably from climate change as well."

"And water? What can we say about water?"

"What about the water in Fiji?"

"Water is good. It tastes—"

"No. Actually we've heard some complaints about it."

"And we had a drought. Recently we had a drought."

"Drought and then heavy rain all of the sudden. And then it just became really cold."

"The Rewa River has become really salty and they've actually found sharks."

"What? They have?"

"They have found sharks."

"Wow."

"Yeah. What else?"

"Well, I guess of all the Pacific Islands Fiji is pretty okay."

"Because it's a volcanic island. It's a volcano. But Kiribati and all that. It's really bad in Tuvalu."

"Yeah. We haven't had a great impact."

"Does climate change affect, you know how we have our cyclone season? Have you guys noticed how every year it becomes stronger? Each year it becomes worse in the flooding and damage being done."

"As small Pacific Island countries, it's hard, you know? Because we don't have all the resources and the money to get back on track. So it takes a long time."

"So we just depend on the weather."

"That's pretty much all we know about climate change and water."

"So please do come to Fiji if you're hearing us."

"It's a nice place."

"Lovely place."

"Way better than other Pacific Islands."

"It's better than anywhere in the world. The best country to be in. Don't you find Fiji great?"

"Friendly people. Beautiful people. Hot guys."

"But it doesn't beat the guys in Samoa."

After turning off the mic, I had an animated conversation about *RuPaul's Drag Race* with the student next to me. I passed the unwritten test of coolness and was invited back to the table the next day.

~

It was raining in the way that Suva does, thick and whole and unapologetically drenching your everything. I found the hut where my new group of friends had relocated and told my new friend to close his eyes and put out his hands. He made a show of it, "Mm-hmm, I love surprises!" I pulled out a tube of silver sparkly Sally Hansen nail polish from my backpack and put it in his hands.

"Oh, *girl*," he said. "I am going to treasure this forever and for always. I barely want to use it, but then again I really do." He set about the business of painting everyone's nails at the table. "Hold out your hands," he commanded. "Darling, you look fabulous."

A new friend joined us, one I hadn't met the day before. She was butch in the conventional sense, a leaned-back swagger in her step, her clothes baggy and masculine, hair short and undercut on one side with a shaved part like the boys.

The guy cupped his hands to his mouth and whispers to me: "She's a lesbian."

I whispered back, unfazed: "Me too."

The shock registered on his face, familiar. I make an effort to conceal my sexuality while I am traveling, and apparently I do a pretty good job of it.

Later that afternoon we braved the rain to take the local bus into town for lunch of white rice and stir-fried veggies and fish.

I wasn't expecting to see lesbian drama in my first week in Fiji (or at all, for that matter), but there it was, like the ocean, waiting—unconcerned with my existence and yet completely immersive.

It was when we came back that it happened. We had just ducked into the cover of a rain-free awning near the university's front gate. The lesbian shook the water out of her eyes to take stock of the scene.

A long-haired girl sat with her back to us, one leg over the seat of the bench, face-to-face with a guy in cut-off sleeves. She stroked his bicep, swiveling her hips on the wood.

"I could punch him," the woman next to me said, a fist forming in her hand.

"Don't do it, girl," her friend said. "He's not worth it." But she didn't hear or didn't want to. She walked over with her head held high. The guy on the bench saw her approaching and said something inaudible and offensive. The lesbian made true to her promise. She threw the first punch. He came at her with both fists flailing, windmills of rage. With one duck and swerve, she came out unscathed.

Her friend covered his eyes. "I can't look. Those two have been on and off since high school. It's a big mess."

I nodded, knowing that this kind of mess has no borders. We could be anywhere. The girl they were fighting over sat on a bench with her head between her hands, covering her ears.

COCONUT SHELL

In Suva, I continued to walk and to listen, and to recount my stories to Esther each evening over bowls of rice at her family's dinner table. On one particularly hot afternoon, I met a man named David who was selling coconuts on the side of the road. In an attempt to cool down, I paused for a drink. The cardboard sign—that day flipped to "tell me a story about climate change"—announced my intention.

"This is Raiwaqa, the most dangerous place in Suva," David told me. "I'll tell you a story about climate change." He used a machete to hack off the top of a green coconut with a few deft swings. On hot days, he told me, wiping his brow, the coconuts sell quickly. But on cooler days, his sales lag. "We coconut sellers depend on the weather."

David pointed to the traffic passing by. "Climate change happens here because of lots of smog going around here, the cars," he said. "People burn their rubbish."

The best water in Raiwaqa, David declared, comes from coconuts. "As far as we know, this is medicine," he added. "The water is very sweet."

The water "is not good here, because we're not living in the States where water is very clean and nice." I thought about Flint, Michigan, and all the EPA superfund sites where my own country has failed to deliver safe water to its people. But I didn't say anything. At that point, in 2014, I was too hesitant to intervene. My mouth was full and empty at once.

I finished sipping from the opening at the top of the coconut. David used another long knife to crack open the shell and handed me a piece I could use to carve out the slim layer of flesh, the story within the story.

Two days later, I walked through Raiwaqa again. I didn't see David; his coconut stand must have moved. Another man, Joss, noticed my sign and walked up to introduce himself. He told me that he used to go fishing at Suva Point, a place where, he said, American soldiers piled up rocks during the Second World War.

As a kid, he walked to the end of those rocks to go fishing. "But now," Joss told me, "I've seen some of those rocks are covered with water. It never used to be like that before. The water has been rising."

Now, to get to his favorite fishing spot, he has to swim.

GOVERNANCE

Rhonda Robinson works as the deputy director of the Water and Sanitation Programme at the Secretariat of the Pacific Community—a regional organization that supports twenty-two countries across the Pacific, spanning from Palau to the Federated States of Micronesia to the Republic of the Marshall Islands to Fiji. I think of her as the guardian of the blue. A friend who was following my travels online recommended that I reach out to Rhonda directly to ask for an interview. To my surprise, she said yes.

I walked to Rhonda's offices in Mead Road, a suburb of Suva on a hill above the rest of the city. Palm trees in terra-cotta pots lined the end of the driveway. Above an aqua-trimmed roof, a Fijian flag ruffled slightly in the breeze.

Rhonda invited me to sit down in her office under the steady motion of a ceiling fan. We cradled two cups of tea. She told me that her main role is to support the national governments in the Pacific to manage their water sustainably. This involves "really understanding what the needs of the countries are and responding to those needs the best we can," she said.

"In the Pacific, communities and households use a combination of water sources from rainwater to groundwater to surface water," Rhonda told me. Her team helps governments measure their water resources, and then analyze and monitor that information to support decision-making. In other words, she considers the whole water management infrastructure, from where it leaves the resource in pipes and pumps to where it reaches the consumer.

In the past, water governance used to be about just getting pipes and pumps in. Then "there was a realization that that alone doesn't help people get water at the end of the tap in a safe way," she said.

Now, water governance is about strengthening policies, legislation, and consultation processes. Because much of water management happens at a household or community scale in the Pacific, this means engaging directly with the people in their communities who are overseeing water resources, supplies, and sanitation.

Rhonda mentioned the Millennium Development Goals as a metric—a list of eight goals about poverty, hunger, disease, illiteracy, environmental degradation, and discrimination against women that UN member states had agreed in 2000 to try to achieve by 2015.[1]

"In terms of the progress we've made against the Millennium Development Goals, the Pacific hasn't done so well," Rhonda said. "And we've done even worse in the area of sanitation than we have in water."

In Tuvalu, the Secretariat of the Pacific Community helped the government analyze the storage capacity of their existing rainwater harvesting systems and make decisions on which infrastructure investments to make moving forward: more gutters, rainwater tanks, or repairs. In Kiribati, two projects aimed to improve water supply systems by monitoring groundwater. In Fiji, a project near the airport in Nadi strengthened the community's ability to respond to floods.

"There's a lot of work that needs to be done," she said. "But there are a lot of committed people, both here in the region as well as within the governments and in the community. I think we just keep working at it."

LEAKS IN THE DIVINE

Walking down the hillside from Rhonda's office on Mead Road, I met Bill, an employee from the Water Authority of Fiji. The Water Authority was created in 2007 to manage water and sewage systems in the country. Bill and his colleagues were reading and disconnecting water meters.

"Water is from heaven," Bill told me—simple, elegant. "Without water we cannot live."

"My point of view is that water, when I mention water regarding climate change—" He paused. "It's up to God. Whatever happens, we should thank Him," Bill said. "We'll go through struggles and changes," he added. "It's all written in the Bible. We should expect this, if you believe in the Lord Jesus Christ. I am a Christian. That's all I know about water. Thank you so much."

Bill turned back to reading the meters. A colleague read out numbers in the cadence of documentation: seven, four, two, seven, three. Bill wrote them down on his clipboard. I turned my microphone toward the hiss of the pipe, listening to the presence and the absence of flow.

"You can see the leakage," Bill said.

He tried to convince one of his colleagues to speak with me, but they were too focused on the task at hand.

BITE

Not all of the stories I heard were biblical.

Willie and I spoke in Lami, a town northwest of Suva, in the patches of sunlight on his porch. Earlier that day, Willie had introduced me to Milo, a malted chocolate powder that, when mixed with hot water, creates a cup of sweet, chocolatey comfort. Willie was a bit younger than me with a head of tight curly hair. He advised me to put coconut oil on my own, which left it shiny and slick.

The details of how we met blur now. After a certain amount of time, the chain reaction of hospitality kicked in. I met one person, who told me a story about water, and then was so enthusiastic about my project that they insisted on setting me up with the family or friends the next town over. A few text messages later, I would have a place to sleep for the next few nights. And even as I ventured farther afield, first on buses and then on boats, I returned frequently to Esther and her family in Suva. Esther insisted on giving me her own bed and sleeping in the front room, an act of kindness that I hope to pay forward to other travelers in the future.

At Willie's family home, away from the traffic and bustle of the capital city, things slowed down. We spent long hours on the front porch watching music videos or napping the afternoon heat away. A cat in his neighborhood had given birth a few weeks before, and we passed the kittens, orange and squinting, between us. After the heat broke that evening with a stream of rain, Willie was ready to tell me his water story.

In Fijian mythology "Dakuwaqa," he told me, "is a shark. They call him a *vu*, so it's like a god. And he wanted to prove to all the other gods in the sea that he was the strongest."

Dakuwaqa went to Rewa, the province on the island of Viti Levu, where he met another *vu* who took the form of an eel.

"Dakuwaqa asked for a challenge," Willie told me. "He said, 'I am Dakuwaqa and I am the king of the sea and I am here to prove that I am the most powerful god.'"

They fought. The eel died. Dakuwaqa continued on to Kadavu, an island about sixty miles south of Suva. "The way to get there is through a passage where the gods can go," Willie told me.

Dakuwaqa didn't know that the passage was guarded by the *vu* of Kadavu, Rokobakaniceva, an octopus.

The shark woke up the octopus. "He asked: 'I will challenge you so that I can go through that passage and go to another village,'" Willie said.

They fought. Rokobakaniceva wrapped four of his tentacles around Dakuwaqa's abdomen, tail, and mouth. In the other four, he held rocks. "He asked Dakuwaqa: 'If you want to leave, tell me now.' And so Dakuwaqa said yes. 'Then you have to promise me not to bite any people from Kadavu and never come past this passage again.'

"And so Dakuwaqa left back to his village on Taveuni and he stayed there," Willie concluded. "And that's the reason why the people from Kadavu don't get bitten by sharks."

TRADITION

Back in Suva, Vilisi and Analisa, two NGO workers, stopped me on the road. We had been walking in opposite directions. After reading my sign, they turned around to flag me down.

"Can we speak with you?" they asked.

"We saw your label here saying tell me a story about water," Analisa said, laughing. "We thought, *Hey, why not just go ahead and find out from her what she's interested in?*"

Analisa told me that she comes from Nasama in the Naitasiri province of Viti Levu, about thirty-five kilometers from Suva City.

"The community needs a good source of clean water," she said. There are fifteen families in the settlement, each with six or seven people. People are still drinking from the well, she told me, but the water is unreliable. Rainwater collection tanks would be an improvement, but they are expensive.

"In this kind of weather they'll be happy because there will be water in the tank. When it is dry, they travel about one kilometer to the nearest big creek to bathe," Analisa said.

"We've approached the Water Authority and government departments. They've done a feasibility study. We've got everything and it costs about a hundred thousand [Fijian dollars] and nobody would fund it," she explained. "One or two years ago we received water tanks; these were two-hundred-liter water tanks from the Direct Aid Program, that's Australian aid. That's two hundred liters for a family of six, seven. So people are still drinking from the well, washing and cooking. We are trying to find ways to improve."

Vilisi chimed in. "My community is a village community far away from the main island, Viti Levu. We have to travel by boat for one night and half a day. The island is called Taveuni. My village is Weilagi Village. And with Weilagi Village they face problems in regards to clean drinking water," she said. "There is not enough."

The current reservoir, a concrete pool on a hillside, Vilisi told me, was made over fifty years ago. Since then, it hasn't been developed. Over time the concrete has degraded. "They're looking for ways to upgrade their current water source," she said.

Vilisi told me that when she went to do a needs assessment survey, families in Weilagi told her that "sometimes they'd see insects coming out of the taps, or tadpoles"—here, Analisa laughed—"or the children would sometimes suffer from diarrhea or skin diseases because they're accessing dirty water."

"Lately we've been trying to look for ways to help them get access to clean water, getting linked to other NGOs or international organizations that do water projects," Vilisi said.

In wet weather, in king tides, or in heavy rains from a cyclone, water can come into the village. Half of Weilagi floods, Vilisi told me, and families have to move to high ground.

"They don't have an evacuation center," she added. "They don't really have the income or finances to move their homes higher up or on piles. . . . So it's really hard."

This thought—looking both forward and backward into the water-related history of her community—sparked a reflection on fish. "I was talking to a couple of old people in my village," Vilisi went on, "and they said that before they used to fish a lot, but now there is no more fishing because the fish don't come close to shore anymore."

Climate change, she noted, has changed the way that traditional knowledge is shared in the community. "I think that links to climate change: changes in temperatures and change in the water temperature," she said. Now that the smaller fish have moved elsewhere, women no longer practice the traditional method. "And the young people don't really care because the older women are not really teaching them anything anymore. Most of the young people have moved to urban areas to find work or focus on studies and other things," Vilisi said, implicitly counting herself among this group. "The loss of traditional knowledge can be linked to climate change as well."

But there are some areas where traditional knowledge in Weilagi is having a revival. Most people in the village farm *yaqona*, a pepper plant whose roots are used to make the brownish drink grog.

"Now they are practicing mixed methods of farming," Vilisi explained. "Maybe it's an overuse of chemicals, but they found that *yaqona* doesn't grow like it used to." To solve this, people are mixing *yaqona*, taro, and other plants side by side so that the soil retains its nutrients.

Vilisi gestured out to the sea beyond Suva, which I couldn't see, but could smell at our backs—the salt that unifies everything. "Interestingly, when you're doing this, you're talking about water, linking it to climate change," she continued, "Right now they have *Uto Ni Yalo*, the canoes."

In September 2014, four *vakas*, or voyaging canoes, set sail from the Cook Islands, Samoa, Fiji, and Vanuatu—all headed to Sydney Harbor in Australia in time for the November 12, 2014 opening of the IUCN World Parks Congress.[2]

Together, the Mua Voyage was bringing back traditional sailing, Vilisi told me. Each *vaka* was led by an Indigenous crew.[3]

"I went to see the canoes before they left." Vilisi exhaled, then smiled. She struck up a conversation with a woman who was sailing from the Cook Islands. "She mentioned that near her home, near where she lives in the Cook Islands, they're trying to retain the corals. The coral bleaching has been too much," Vilisi said, "so the fish as well are not coming closer to shore like before.

"Right now, they're trying to create awareness about that and doing coral farming, things like that to retain a marine environment. And she said it's not like before. It's changed. It's changed a lot."

I walked downhill toward the harbor, thinking about the wind.

MAP

One afternoon, I bought a map made by the Department of Lands and Surveys. There are over three hundred islands in Fiji. The piece of paper showed me the fist of Viti Levu, the main island, with Vanua Levu, a jagged stripe, above. I traced my finger over the Pacific—salty blue—with Fiji's tinier islands strung out like dice and poker chips on a card table. I started to annotate the ink blots, circling the names of the places people told me to visit.

Inside the map shop, a man named Meli told me about his home.

"I hail from the island of Gau in the Lomaiviti Province. My story is just very short concerning the village that I come from, Nukuloa. It's right there," he said, pointing to Gau, an ear-shaped island east of Suva. Nukuloa, he told me, is renowned for its freshwater spring.

"Our water is, what would you say, it comes from the bottom of the rock, eh? So if you take a jug or a bottle and you fill the water from the tap in the village, if you leave it for a while you'll see the jug and the bottle sweating," he said. "The water is really cold and very fresh."

Meli told me that the village leaves the water running—the flow is simply too strong otherwise. When a visitor comes, the first night they sleep, they hear water running, they'll go and turn it off, close the tap. But it's just something that we let the taps run the whole day. It's just because of the pressure from where the source is coming from," he said.

"If you go and see where the source is, we have a pool." Meli told me that I might expect to see water bubbling there, but it's stagnant. "If you look closely to the bottom of the rock," he said, "you'll see small bubbles, tiny bubbles, coming from the source."

"But that's basically the water story from Nukuloa-i-Gao. Thank you. Over to you, Graham," he said, miming the voice of a TV presenter. Everyone in the map shop laughed.

BLACK SAND

After this story, an invitation: "You have to come to Nukuloa," Meli told me, tracing his finger across the map's ocean, roughly sixty miles east of Suva. "We have the best spring."

That is how I found myself, five days later, on a fifteen-foot fiberglass boat with an outboard motor, my knees hugged up around jerricans of gasoline—sea salt crusted on my face and clothes—bouncing across a turbulent patch of sea.

East of Viti Levu, the island of Gau rises up out of the Pacific—small, lush, hilly. Blue and yellow houses dot the shore. Strings of laundry spanned the spaces between. The harbor was made accessible by the high tide. Around three thousand people live here.

We unloaded the gasoline first, then ourselves. I learned that night that there was no electricity on the island—the jerricans were to supply the generators that purred behind a few well-off houses, and only when necessary. When one household ran their generator, people came over to charge their phones from a power bank, or maybe watch a movie. The limited electricity was for the whole community to enjoy.

The word *nukuloa* translates to "black sand." The spring-fed water is so fresh in Nukuloa that, legend has it, ancestors of this village stole their water from another island. While being chased by the other islanders, the ancestors threw the water at Gau, forming the river that runs along the top of the island like a thin blue braid. The water is a source of pride for those who live there.

PLASTIC

Two pathways lead up the hillside from Nukuloa: one to the spring, which is a sacred site. Natanieli Seru, the kingmaker of Nukuloa (the leader of his clan), gave me permission to visit the source. We walked up the hillside together and stopped in front of a concrete rectangular pool with minuscule bubbles beneath the surface. A leaf floated on the water's skin.

"It comes from under this big mountain here," he said, "and the main stream is far away on the other side."

Natanieli told me that years ago, a bottled water company approached him to ask to bottle and sell the water in exchange for electricity on their island.

"They want to convert this into a business," Natanieli said, looking at his reflection in the spring. "Problem, eh? What does business matter? For money and things like that. But this is God-gifted to us. Free. It comes out from under the earth, and it's well filtered by the soil. And see the gravel, see the calmness of the water? You see the surface is very clear?" he asked.

"I was thinking about the villagers. Will it be good to them? Money is not everything. This is the gift of God. This is a divine gift from above. You don't have to pay for anything. It's overflowing," Natanieli explained. "They never run out of water at any time. It's always like this."

The elders gathered under a thatched hut to discuss the issue for days. Eventually, they voted against it, believing that the water would be a more valuable asset if kept for their village rather than shipped out to the rest of the world.

So, for now, Nukuloa still uses diesel generators. At night, it was the darkest dark I had ever seen. People who have lived there all their lives navigate the pathways barefoot, by feel.

A second pathway above Nukuloa leads to a cool waterfall, a favorite afternoon swimming spot for the village children. I joined a group one afternoon. We took turns jumping off the surrounding boulders into the deep pool below. We played tag. My ears filled with water.

GIVE & TAKE

I dug my elbows into her shoulders and the space between her clavicle, focusing the whole of my energy on relieving the stress. The lavender oil slacked between my fingers and her skin.

Suzanna's sixty-year-old back was tight from caring for her eighty-year-old mother, who is bound to a wheelchair in Gau. "I lift her from the bed into the chair, from the chair to the toilet, and back all over again," she sighed. "It's a lot."

Suzanna lay out on top of my Therm-a-Rest sleeping pad and exhaled "*nn-nn*" when I pushed my thumbs on the knot directly.

"Is this okay?" I kneaded and pressed until something in her loosened.

My mother taught me to give massages when I was eight years old. I remember practicing for the first time on my grandmother, how soft her face looked after even the simplest of back rubs. Since then, massage has been one of my favorite gifts to give—all I need is my hands and a willing recipient (oil is a luxury).

Ili was next. She took off her shirt and lay facedown, tilting her neck to the right. I unhooked the four clasps of her bra. Ili has three tattoos on her upper back: a skull with two flowers coming out of the ear, the word LOVE, and a line about an inch long.

After the massage, Ili took my hand. "You come?"

I nodded and shrugged a *sulu* around my hips. We followed the footpath to the next village over, Levuka-i-Gau, her home. Mango seeds littered the path between. She pointed up to the mango trees where the fruit was just starting to get ripe.

"Not yet," she said. "Soon."

We passed two bulls grazing in the tall grass. I put my pointer fingers on either side of my head, bowed my neck down low, and stuck out my rear, imitating the bull's proud stride, defraying any tension by making fun of myself.

"*Eeeo!* Yes!" Ili shouted, exuberant.

I swiveled my hips, stopped walking, backed it up, got low, popped up, vogued, freestyled with my wrists arching high to the sky, and threw in a few rib isolations for good measure. Ili laughed her tall laugh and joined in. "Hop hop! We hop hop!"

Dancing without inhibition has gotten me far while traveling. What I can't communicate verbally, I can say with my body. Plus, it just feels so good to move.

I followed Ili off the main path as she took my hand and led me into her village. Bright roosters and peeping chicks pecked the ground. The passage snaked between a cluster of tin houses. We wound our way beyond the seawall where children splashed and swam in high tide.

Ili's husband chopped wood. She greeted him with a warm smile and insisted that he put his ax down for a moment to shake my hand. The smell of woodsmoke and curry wafted from a fire and the accompanying big pot. I smiled as I inhaled. Ili reached for my hand and we walked farther up the hill to meet Suzanna's mother. Suzanna wheeled her into the main room of the house.

"Take off your shoes," Ili reminded me. "*Tombe.* Sit."

I took a seat on a squishy foam couch against the wall. Suzanna pushed the chair over a woven mat and as close as the wheels would come. Her mother looked me in the eye and said a sentence in Fijian. I looked to Suzanna, the question in my glance.

"She says she can't walk."

I stood up to touch the woman's shoulder, saying with my body what I am unable to say in words. The radio hummed away in Fijian, the rhythm of the language both familiar and strange.

"We go?" Ili said.

I thanked Suzanna and her mother—*"Vinaka vaka levu!"*—and followed Ili out the door and up a hill. We passed a group of Fijian men who leaned against the trunks of coconut trees, passing a cigarette between their lips. Ili guided me into the dark commissary where she introduced me to Lucy. It took a moment for my eyes to adjust from the brightness of outside. We exchanged a handshake and a kiss on the cheek. Then Lucy busied herself putting things into a plastic shopping bag: a package of butter biscuits and tropical juice mix.

With my eyes finally acclimated, I looked around to take stock of the room. A golden-lettered Santa Claus poster wished people who were exiting a Merry Christmas. Two pairs of cleats sat lonely on a high shelf, gathering dust. Boxes on boxes of chicken-flavored ramen noodles leaned on each other in the room's far corner. There was a mug full of packaged toothbrushes near the cash register, the used versions of which I had seen littering the beach at low tide. Lucy handed the shopping bag to me.

"I didn't bring any money—" I protested. But Lucy waved her hands nonchalantly, insisting that I take the biscuits and tropical drink mix. I bowed my head in thanks.

I followed Lucy and Ili into an adjacent room with woven mats on the floor. On the wall was a poster of a blond Eve reaching for the forbidden fruit and a shelf with seashells of various sizes and colors. Lucy called her daughter in from the kitchen and asked her something in rapid-fire Fijian.

Lucy took out a sewing box and unraveled a spool of brown thread. In one swift motion she threaded the needle's eye. Lucy's daughter returned minutes later with two handfuls of yellow flowers. She took a moment to look at me from head to toe, and then sat down next to her mother, biting off the end of the flowers and handing them to Lucy, who threaded them onto the string, one by one.

"*Bua*. That's how we call the flowers," Ili explained, lifting the completed necklace up and over my head. "It smells good, *mm*?"

I took a deep breath and closed my eyes. I smelled sweet. Ili put one of the *bua* flowers behind my ear, too, for good measure. I felt gorgeous.

Lucy called her eldest son in from the yard where he was smoking and gave him a short sentence of instruction. He jogged out of the room and reentered a few minutes later with a coconut that he'd climbed up a tree to fetch. Lucy's daughter went to the kitchen to find a sharp knife and hacked off the coconut's top in three movements. She handed me the coconut to drink.

Ili and I danced the whole way home. Such grace, the act of giving. Hands to hands.

CHRISTMAS ISLAND

On one of my last afternoons in Nukuloa, I pulled out a pack of crayons and a few pieces of paper from my notebook and sat under the shade of a tree to draw. Next door, a family was pit-roasting a pig. Three children, Ledua, Villy, and Laisa, walked over to ask what I was doing. We took turns drawing each other in crayon and pen, laughing at the results: boxy teeth, lopsided eyes, out-of-proportion arms. I pressed too hard on the red crayon and it broke in half.

A few hours later, Natanieli Seru stopped to say goodbye. "I have another story to tell you," he said. We walked over to his front porch and sat down on the steps. Though his story took place in 1958, it haunted me for months. I didn't have my audio-recording equipment with me—only my notebook. I scribbled his story next to the clumsy portraits I had drawn that afternoon.

"They told us to turn our backs. We stood on the beach, all three hundred of us from the Fijian Navy Reserves. It was a beautiful, blue day on Christmas Island, the day they tested the British hydrogen bomb. I could feel the heat. The top of the explosion was like ice cream. Billowing fire," Natanieli recalled.

"Some of us, when we came back, our hair fell out. Most of us, when we came back, had miscarriages. I knew something was wrong when albatross fell out of the air, dead. Birds farther away flew into buildings, blind. My son was born two years later. He was always sick."

Even then, in 1958, the Pacific Islands were used as the testing grounds for something dangerous and unknown. These pinpricked edges in the middle of the ocean, the seemingly far-flung places, were deemed disposable, their futures cast aside. That idea, so familiar in a changing climate, is not new.

Natanieli's son, Sakiusa, died at age forty-nine. In the autopsy they couldn't find anything wrong.

LOW TIDE

I took an overnight ferry to Savusavu, a town on the south coast of Vanua Levu island, where I stayed with Sesa, the aunt of a storyteller I had interviewed in Suva. Sesa's wooden house was propped up a few feet above the ground, just in case the high waters came. We walked down the road to the town's hot spring to breathe in the sulfuric evidence of volcanic activity and to marvel in the heat. During the day, Sesa taught me to make roti using flour, water, and soybean oil. In the afternoons, when the light was long, she took me to the volleyball court in the center of her community. We played with *sulus* wrapped around our waists. Sesa, on the court, was a shark.

At night, walking to the outhouse, I saw the Southern Cross for the first time: a constellation stitched with stories. The way we talk about climate change is often numerical and abstract. It's more difficult to ignore changes we can see. What if we could paint the CO_2—make each molecule visible in the sky? Would that change the conversation?

After a week in Savusavu, the neighborhood threw me a farewell grog party with fresh fruit. A trio sang accompanied by the guitar and a three-stringed ukulele. One of Sesa's nephews climbed a mango tree. "Water is everywhere! In fruit, even!" he said, tossing the fruit down. We picked it up as it hit the ground and peeled each mango to eat the stringy fibers. Pieces stuck between my teeth. Juice rolled down my elbows.

Another man doled out drinks of grog from a coconut shell. I had a "low tide" (very little) while he and the others indulged in a "high tide" (full cup).

"*Vinaka na sere*," I said, testing out my nascent Fijian. The juicy vowels. Thank you for the song.

Earlier that day I stopped by the water to sit in the arm of a palm tree and watched the beginnings of a sunset. I took out my notebook to write. The low-lying clouds were moving in one direction and the upper-level clouds another. The thing about climate change is you can't paint it. It is chaos. It is motion. And so, in this journey, am I.

MANGO SEASON

Devorah is one inch shorter and three years older than me. I could be her if not for an accident of birth.

We sat at a picnic table in Labasa, a town on the northern side of Vanua Levu, on a hot November afternoon, sharing a plate of french fries with barbecue sauce. I had taken the bus north from Savusavu in order to listen.

"How did you meet Isoa?" I asked.

"I was seventeen and at secondary and he was twenty-two and at university in Suva," Devorah sighed. "I got pregnant and my parents said it wouldn't look good to have a baby out of wedlock." Her father is a preacher at the Methodist church.

"So here I am, three kids later." She shrugged and sopped up some of the barbecue sauce with her french fry.

Devorah dreamed of going back to Suva. "Everything is better there," she smiled. "I would live with my family."

We walked through light rain, veering off the main road and through a path that bisected a sugarcane farm studded with bleating goats. My shirt started to soak through to the skin.

Johnny, age two, and Isimeli, who just graduated from kindergarten, were in tow.

"The rain is good for the mangos," Devorah explained. "We had three months of no rain and everything turned brown. No fruit. But now the rain brings bounty."

Devorah pointed to an overgrown mango grove just off the path. We slid through slick mud and up the small hill to gather fruit there. Johnny and Isimeli threw sticks up at the branches to see what would fall. We took the easy fruit first. Then Devorah scaled a branch and tossed mangos down to me, one by one. The fruit we gathered overfilled and threatened to tear two plastic bags.

"Try one!" she said. "Take a drink."

The orange pulp made a mess of the front of my shirt. I reached for a second, and a third.

"This kind of mango is good," Devorah said, taking a bite and ripping the skin with her teeth, "because the seed is small and the flesh is soft and sweet."

I hadn't known that there was more than one variety of mango. But of course there are. I felt very small (and full of fruit). I will never know the world in all its juicy whole.

Our conversation turned to plans for tomorrow night. Earlier in the day we had run into Devorah's cousin, who invited us to go out dancing in town.

"Come with us!" I pleaded. "I know you're a good dancer."

Devorah laughed. "I would love to, but I'm not free!" She gestured to the kids. Isimeli threw rocks against a tree, and Johnny had gotten his flip-flop stuck in a patch of mud.

~

Before bedtime I used to ask my mother: "How did I get to be your child?"

"You chose me," she said, never missing a beat. "You were floating out there somewhere and had your choice of so many mothers, so many lives. And I feel lucky that you chose me." She punctuated that last sentence with a hug.

Though I doubt it's true, one thing is for certain: if I were born Fijian, there is no way I would be traveling, living this journey.

Financial matters aside (thank you, Harvard funding gods), gender, and culture are their own limitations. More than once, a man lectured me over a plate of cassava and curry: "I just don't like the idea of you, a girl, being out there all alone. Anything could happen. I wouldn't let my daughter do what you're doing. It is not suitable for Fijian women."

I travel not only for myself, but for all the women who are unable to do so.

BAMBOO

At the bus stop in Savusavu, Taniela Nasi told me the early settlers to the area arrived on a bamboo raft from Viti Levu. But they forgot to bring hot water.

"The chief of the village sent three people to go back to Viti Levu to bring their hot water," Taniela told me. "They swam back to Viti Levu and brought the water in the bamboo shell."

When they passed what is now Cousteau Resort, near Lesiaceva Point, the hot water started to fall out of the bamboo. The three people got out of the water, took all of their bamboo, and ran along the beach.

When they arrived at their village site, the chief picked up the bamboo, broke it in half, and spilled the hot water on the ground. "If you go there today, the hot water still remains there," Taniela said. A hot spring bubbles at the site.

"That's why they call this creek Nakama Creek. In Fijian, *kama* is 'heat,' because of the hot water that was brought across the sea from Viti Levu," Taniela said. The hot spring is used for cooking.

When the chief broke the bamboo, he took one piece and threw it across the water to Nawi Island. Today, Taniela told me, there is still a bamboo tree in that spot with only one plant.

SINEW

From Vanua Levu, I took a bumpy ferry to Taveuni. At noon, I was picked up by Eleni Siteri, the friend of a friend I made in Savusavu. Her aunt was celebrating her fifty-fifth birthday that night. We walked uphill from the dock to the party. There was no electricity; a generator powered two big speakers playing Fijian music.

A feast was being prepared over an open fire. The kids ate first. Then the adults sat in a circle to bless the food, and the aunt's daughter gave a speech in Fijian about how grateful she was for her mother. Everyone cried. We ate a fish with blue bones and *rourou* in coconut milk (a green leafy dish) and taro and cassava, root staples. The cake was like banana bread with icing on top.

Later, when it rained, I lay down on my stomach on the floor of a house on a hill that looked out over the whole fabric of the water. A neighbor taught me how to pound out paper from the inside of a tree. We painted a turtle in black ink.

THE CURE

At the center of Wairiki, a village on the island of Taveuni, is a church. Successive waves of missionary activity in the South Pacific, which accelerated in the nineteenth century, mean that Christianity weaves into Fijian stories about water and climate change—specific springs, fountains, and sites of healing have layers, some of which are religious.

Eleni walked me to the Catholic church to stand in front of Mary's fountain: a white shell with stagnant water at the bottom.

"We Catholics believe that whenever we are sick, or whenever someone is down, we come to this fountain and then we drink or wash our face," Eleni said. "If there's an infection, we wash our face here. Whatever we need, or whatever we prefer that's hard to get, we come to this fountain."

PURPLE IS FOR MOURNING

The air was thick with grief. White, sheer fabric hung on every wall, covering even the windows. Four silky blue sheets knotted in bows accented each corner. I could hear women wailing from outside. I shook off my shoes and followed my host, Eleni, into the stream of extended family members who walked toward the casket to pay their last respects. I cast my eyes down out of respect. We moved forward slowly, shuffling. I gulped and found it hard to breathe, as if the air was just out of reach.

When we reached the far end of the room, Eleni dropped to her knees and crossed herself. One of her aunts grabbed the casket like it was the only object tethering us all to the ground. The women around us rocked back and forth on crossed legs, tears flowing freely.

Eleni bent over the coffin and kissed her grandmother's cheek. She dabbed her red eyes with a lime-green handkerchief, turned around, and waved me forward.

It was my turn.

I followed Eleni's example and fell to my knees, trying to distance myself from fear. I had never seen a dead body before. I looked down into the woman's face. She was pale but serene. Tight curls of gray hair wound close to her skull. I wondered who washed her skin, who arranged her hair. Bits of cotton were stuffed in each nostril. Only her head was visible; someone had tucked in a length of white fabric just below her head, obscuring her body from the neck down.

"Kiss my mommy," Eleni's mother whispered from next to me. She smiled through her tears, offering permission. It was a command, not a suggestion. There were others in line behind me, waiting to see their loved one.

I leaned over and placed my lips on her forehead, gently. Something in the temperature of her skin brought me back into myself. Her face was clammy and cold, starting to thaw after two days in the hospital mortuary. In a few hours she would be underground. Eleni's grandmother had twelve children, though I never learned her name. At a Fijian noontime mass, we blessed her body with song. I hummed along, trying to blend my voice with those around me.

As the mass adjourned, a group of holy men dressed in white wrapped her casket in a woven mat and carried her body toward the sea. The whole tide of us followed in procession—singing and strumming and kicking stones on the path—until we reached a plot of land dotted with graves at the crest of a small hill. A fresh, deep hole marked the earth. We gathered there.

The pastor blessed her body as the men lowered her into the ground. Those closest tossed in handfuls of dirt. One man in rubber boots hopped into the grave to stamp his feet on top of the freshly turned, reddish earth, compressing. The ground was damp from five days of rain. I closed my eyes and inhaled the not-so-distant saltiness of the Pacific. When the level of soil on top of her casket reached that of the ground surrounding it, a second wave of men hauled in a pile of stones and set about building a rectangular rock wall around the new grave, about two feet tall. Dirt filled this new emptiness. In a third wave, three men appeared with bags full of coral, which they poured on top of the raised bed. The sound of coral on coral was like glass breaking.

Eleni's mother draped a piece of fabric on top of the coral. Her sisters came to thread fabric streamers around the four tree-branch stakes that marked each corner of the grave. Eleni handed her four-year-old son a wreath of purple leaves with green undersides that she'd gathered from her village last night before sunset. "Purple is for mourning," she told me in a soft voice, her eyes still red. Her son leaped over to join the other children who placed wreath after wreath of flowers on top of it all. The result looked, to me, like a ship.

TAGIMOUCIA

Legends dot the landscape. We walked with dawn at our backs toward the promise of flowers, red flowers laced with a tragic love story. I heard two versions of the same tale told by village elders. In one, a village woman fell in love with a chief, but was forbidden from marrying him by custom. When her wedding day with another man arrived, the woman fled up the hill we climbed with a knife in her hand. People from her village pursued her. At the peak, she stabbed herself. Rather than blood, streams of red flowers flowed from her chest. They bloom every year in December, a reminder of the woman's suffering.

In the second telling, her lover was a man of low status whom her family detested. The two escaped up the hill together. The villagers, noticing their absence, pursued and killed them both as a matter of honor. The lovers' wounds form a bloodred chain of intertwining flowers that bloom on vines every year.

"Horticulturalists from New Zealand and Australia have tried to take specimens of the flower back to their countries to grow," the second storyteller chuckled from his belly, "but all the flowers died. Taveuni is the only place that tagimoucia can survive."

Three of us trudged up the storied hill: myself, Eleni, and a gangly guy in orange basketball shorts. He didn't talk much, and we never learned each other's names. The light was pinkish and fresh in the haze of early morning. With each step I woke up into my body and my voice.

Talking was a good way to keep from thinking about the enormity of the hill that surrounded us. On a particularly steep slope, Eleni turned to me and huffed: "You won't believe this, Devi, but back in 2003 a whole group of kids snuck out of their village at dawn and walked up here alone."

As if on cue, a light rain began to sprinkle our arms. Looking up, I saw nothing but cloud and the long legs of our companion some thirty steps ahead. I gave up much earlier trying to keep up with his exuberant strides.

"I think about them in this rain," Eleni shivered. "They were so young. Ten kids, the oldest not more than eleven years old. They wanted to see the lake for Fiji Day."

Lake Tagimoucia, nestled in volcanic rock, is Taveuni's pride, named for the telltale flowers. Barring a daylong, circuitous bus trip to the north, you can only see the lake's shores from the top of the hills that surround the main harbor. We would have been able to see the lake from the summit of the hill we were climbing, were it not for the thick fog.

"When is Fiji Day?" I asked, eager to change the subject. The forest around us seemed haunted, suddenly, with more than the breath of raindrops.

"September tenth? No, October tenth!"

Our footsteps left squelchy marks in the soil. A few minutes passed without comment.

"But the kids," I had to ask. "What happened to them?"

"My grandmother, the one who just passed away—she was one of the first who found them. Search parties were called off after a week, and it was a month before anyone heard of them. Then one day, a raggedy group of kids stumbled into the village."

We stopped walking for a moment to take a drink of water.

"They were cold. She welcomed the kids into her house and wrapped them up in blankets straight away. When my grandmother called the police to say that ten lost kids had showed up in her village, they thought it was some kind of joke. She had to persuade the officer to come and see for himself. Then she did what Fijian women do best: she gave those ravished children something to eat."

Eleni paused, the tang of the memory clear on her face. "The whole island celebrated their return. Everyone feasted. Our joy changed even the taste of the air."

We had only been walking for an hour, and already my stomach was hollowed out with hunger. Eleni dug in her shoulder bag for a mango and tossed it my way. We continued walking in silence.

"And all the children, they all came back alive?"

"They all came back alive."

Several false summits later, we found the flowers.

The fog, by late morning, was dense and our bodies were coated with a film of rain. I couldn't see more than five feet in front of my face, but somehow the guy with the orange shorts had catapulted himself up into a tree, machete in hand, to bring back a few of the tagimoucia flowers. They were smaller and more stunning than I had expected, each bloom a story.

The farther we climbed, the more the barriers between us disintegrated. The guy with the orange shorts started to open up, though I was still too shy to ask his name.

"See up there?" I followed the line of his finger up to the tip of a cloud where a metal structure barely pointed through the fog.

"That's the Digicel tower. You can't see the Vodafone tower, but it's over there." He tugged a faded navy-blue beanie over his ears and launched into a story.

"The first time I came up here I was a teen. My friends and I woke up and just decided to go. There's a saying among the elders that if you plan to come, it will rain; if you wake up with the idea on your lips, the weather will be good. It was a clear day, windy too. We climbed up the service rungs on the Vodafone tower, took a picture with my friend's cell phone for proof, and walked back to our village by the sea."

I nodded, in awe. After climbing this high, the last thing I could imagine was scaling a cell tower.

"I haven't come up here in a while," he said. "It is a holy place."

SUSTENANCE

On one of my last days in Fiji, I met Marion, a woman selling beaded necklaces at a market in Suva. Seeing my cardboard sign, she smiled and approached me.

"I'm from Nausori," she said, a town about twenty kilometers northeast of where we stood. "I do a little business around Suva and I met this little lady. What's your name again?"

"Devi," I said.

"Devi," she repeated into the mic. "And she's got this thing wrapped around her neck which has a notice board which says, 'tell me a story about climate change.' So it caught my eye. So here we go."

She told me that in Nausori, climate change has affected the rivers. After a dry year, there were two weeks of heavy rain, water that fell "right from the top of the mountains." The high river flooded crops, putting people's livelihoods at risk. In addition to farming, women in her village dive for *kai*, a freshwater mussel. "It's a form of sustenance. They go out and sell that on the side of the road," Marion explained. When heavy rain comes, the *kai* are inaccessible.

The river itself changes with each flood. Marion remembers a time when the river was deeper. Now, soil and runoff from the mountains has made the river more shallow. "Every time we have just a few millimeters of rain, it tends to flood so easily," she said. Life in the river and livelihood for the women living on its banks are intertwined.

3

NEW ZEALAND

GENDER

A man at a bicycle shop in Auckland asked me why I was traveling alone.

"Are you running away from something?" he asked.

"I'm not running away," I told him. "I'm moving toward."

INITIATION

I started riding my bicycle in New Zealand, past the geometric trees and blue volcanoes of Auckland. I had shipped the bike ahead to meet me at the house of two local climate activists, Rachel Dobric and David Tong. They picked me up at the airport and fed me vegan jerky and coffee. I slept on their pull-out couch. David rides fixed-gear bicycles, which he stashed mounted on the wall; I put my bicycle together in his living room and plotted my route. The general directive: south.

Collin Rees, a Harvard rower and climate campaigner who had attended the UN Climate Change Conferences in Warsaw and Lima, e-introduced me not only to David and Rachel, but also to Geoff Keey in New Zealand. As in Fiji, one connection was all I needed to get started. Geoff told me about a climate justice *hui* in Taranaki, a western part of the North Island that juts out like a nose. *Hui* is the Māori word for gathering.

"Sadly I can't make it, but there will be quite a few people it would be good to meet," Geoff wrote on Facebook Messenger. I was on my way.

RING AROUND THE MOON

On the first day, my bicycle broke down south of Auckland, on Highway 22. I pedaled uphill. Tall pine trees exhaled a familiar smell. I was surrounded by farms. A bridge, so small it looked like a toy, beckoned in the distance. Traveling by the power of my own body, I felt free.

Near the hill's crest I heard a metallic ping on the tar, followed by an awful dragging. I stopped to investigate. My rear rack had fallen off entirely. A screw was missing. Other pieces were loose. I kicked myself for not double-checking their tightness.

Two cars passed. The third one, a black car with dents in the doors, pulled over. Out popped an older man with a deep tan.

"Do you need help?" he asked.

"Yes," I said. I told him about the crucial screw that probably fell off kilometers back.

In his back seat, a dog was going nuts; he must not see people very often. I followed the Three Second Rule and made the judgment call to trust this man. He introduced himself as Greg.

I loaded my panniers in the back of Greg's car, and we tightened the back rack enough so that I could ride to his house in Pukekawa. He left me with the address. I counted driveways until I found his. The barking dog gave it away.

Greg showed me around back, and we dug in the toolbox for the right-size screw. Then we fixed the rack next to his paddock of horses. It was six p.m. at this point, and Greg asked where I was planning to ride that day. I was exhausted.

"Is it okay if I camp in your backyard?" I asked.

Greg laughed. "You can camp in my guest room, though you should be aware that the place is dirty," he said. Greg had been living in the house for six months but spent all his time in contractor jobs in Auckland. "I don't have the will to fix the place up," he said. "It was unoccupied for years before I bought it."

Dirty was an understatement. Cobwebs dangled everywhere, even on the glass of the framed photos of his ex-wife and kids. I noted two holes in the ceiling—"Possums fell through last week," he told me.

The dog wouldn't leave me alone and kept trying to crawl into my lap. I summoned the energy to shove him off.

Greg disappeared and returned with a bright neon reflector vest, the kind contractors use, and a map book from the New Zealand Automobile Association. These gifts would quickly prove invaluable. Then he fed me broccoli and beef and baked potatoes. We talked about my route. He warned me about distracted drivers and told me to wear sunscreen. He also said that I shouldn't be doing this alone.

"When I saw you on the road, I didn't know if you were a girl or a guy," he said. "I just wanted to help."

Greg's house was stacked with old newspapers and magazines. I picked up a 1978 directory of thoroughbreds in New Zealand. They had the most wonderful names: Straw Honey and Ring Around the Moon.

Greg worked in construction; he knew about the underbelly of things. The previous week, he had worked on a contract at Waikato Water Treatment Plant in Tuakau, about an hour's drive south of Auckland. This is one of three main sources for Auckland's water.[1]

"They're going to expand the plant because there's not enough water in Auckland to cope with the number of people moving into the city," he said.

In 1996, the census counted 347,768 people in Auckland City.[2] The latest census, 2018, counted 1,571,718 people.[3] That's a lot of water: 160 liters per person per day, by the estimate of Watercare, an Auckland Council organization.[4]

"They take the water out of the Waikato River," Greg said. "It goes through a big plant, and they strain out all the excess. There's a lot of mud in the water. They take all the water off the top of the river, but there's heaps of mud and the mud comes out of one of these treatment plants. Big piles of mud. It's beautiful-looking mud, like topsoil."

As the night wound down, we talked about dreams. Greg told me that if he could go anywhere in the world and do anything, he would tour the United States on a motorbike. He has relatives in Bakersfield, California.

That night, a possum chewed a hole in my food bag and ate a loaf of bread. I left Greg's house a little after nine a.m. He had to call into the room to wake me up because I was so exhausted. My legs hurt. A few kilometers down the road I saw his car pull up and stop in front of me. I had forgotten one of the water bottles that I filled up on his kitchen counter.

This is why I travel—for the moments of being overwhelmed, for the sweetness of learning. To be humbled.

METHANE

At the crest of a hill, I found another biker lying face-up in a patch of shade. He looked like he was sleeping. Two small panniers balanced on his back wheel; he couldn't have been just a day cyclist.

"Hi! Are you all right?"

He sat up, dazed. "Yeah, I'm just taking a rest."

"Are you headed this way or that way?" I pointed.

"This way, to Raglan."

"Me too!"

Derek and I rode together for the rest of the day, egging each other up the hills. Derek worked for the Bank of New Zealand in Auckland and was riding to Raglan and back for the long holiday weekend—it was Auckland's one-hundred-and-seventy-fifth birthday. By the time we pulled into the beach town, it was an hour from sunset. Time to improvise.

I saw a couple walking on the road by the water and asked them where the nearest backpacker hostel was. They pointed the way but said that they were sure that it would be full.

"You could stay with us," they said, introducing themselves as Bill and Julie. "We have an extra bed."

After stopping for pizza, Derek and I pedaled the last kilometer uphill to their home. Over biscuits and tea, Bill and Julie told me that they didn't believe that climate change is "a man-made thing."

"I'm a climate change skeptic," Bill said. "I think we've been conned by whoever is trying to con us for some financial gain."

Bill told me that in 2003, the New Zealand government proposed a tax on all livestock for their methane emissions. "It was nicknamed the fart tax," he laughed.

"—because they emit gases, greenhouse gases," Julie continued.

"—through farts," Bill said, finishing the thought. "And when people heard this was a government policy—"

"It caused protests, didn't it!" Julie said, stifling a laugh.

The Guardian reported at the time that four hundred farmers with twenty tractors gathered outside Parliament to voice their opposition to the tax.[5] Over sixty-four thousand farmers, nearly half of all farmers in the country at the time, signed a petition against the tax. One member of Parliament led a cow named Energy up the steps of the building.

At the time, New Zealand's forty million cattle and sheep produced an estimated 60 percent of New Zealand's greenhouse emissions—more so through burping than through farts—amounting to more than the transport and power industries.[6]

"People were just lying on the ground, laughing their heads off," Bill said. "They couldn't believe that this was going to happen."

Ultimately, it didn't.

Bill and Julie voiced a distrust in the way that television stations communicated news about climate change. "I think the news is completely slanted," Bill said. "Every single news item you get on the TV in New Zealand about so-called climate change is always negative stuff. They always show you icebergs floating in the ocean and cracks in this and that."

"And then they show all the ice cracking off the end of a glacier, but that happens naturally in the summer," Julie added. "So it's hard to know whether it's true or not."

They were both sick of hearing about polar bears. "They pick one isolated incident of this poor little polar bear with two cubs that happens to be on a little ice floe that's drifting away," Bill said, "but it doesn't give you the complete picture," he added. "Something is not right. Some people are not telling the truth."

In the morning, Derek and Julie and Bill and I went for a swim at the beach. The waves were big. The tide was coming in. I followed Bill's lead about when to jump over or dive under their crests. On some, I misjudged and got a little splash of water in my face as the wave started to break. I stayed in the sea until my fingers turned to raisins.

Bodysurfing was a joy, my hair pushed forward into my face by the raw force of the water, its tumbling. I inhaled salt. I could feel it healing the scrapes and nicks on my knees.

~

Ten kilometers outside Raglan, a man at his mailbox watched me pass. "How much is that bike worth?" he asked. I said that I bought it in the States a while earlier and couldn't remember.

"How many gears you got?"

"Three big ones, and plenty of little ones in between."

"Oh really?" He squinted, the wrinkles on his skin pronounced. "Back in my day, bikes didn't have gears. If you wanted to go up a hill, you had to pedal harder."

"Are you a farmer?" I asked. It's the questions that get me places.

"Yes. Dairy. All my life. This is my land." He pointed. "Those are my cows. That is my mountain. I used to cycle up it when I was younger. Well, I did. Now I'm a broken machine. A broken record. I used to pedal up a track that goes right along that ridge." The farmer gestured to the mountain with his arms, his glance tracking up the peak and then back to me and my bicycle.

"I'm going that way."

"You be careful now. The road is unsealed."

I waved goodbye. Hours later, I took a wrong turn, down a street that dead-ended at a waterfall.

PAPATŪĀNUKU AND RANGINUI

I met Lisa Hopa in the parking lot for Wairēinga, or Bridal Veil Falls. She was adjusting a reflective silver accordion over the dashboard of her car.

"Do you know how long it takes to walk to the falls?" I asked.

"Ah, only fifteen minutes maybe."

"Can I join you?"

I introduced myself and my project as I locked up my bike and covered it with a tarp.

"Have you ever met a Māori person?" Lisa asked.

"Not yet."

"I'm Māori," she said. "*Nau mai, haere mai ki Aotearoa*; so, welcome to New Zealand." Lisa told me that she worked as a primary school teacher in Waikato.

We descended the steps to an outlook where we could see the water coming down in pirouettes and crashing into rainbows below. At the bottom of the falls we sat down in the shade for a quick course in Māori mythology, droplets of water just reaching our faces.

"The Māori word for water is called *wai*," Lisa told me, the same as the word in Tuvaluan. "It's used in a lot of ritual rites, or *tikanga*, as we would say."

All inanimate objects, Lisa told me, have a life force and a genealogical tree that can be traced back to Papatūānuku and Ranginui, the earth mother and the sky father. In the beginning, Papatūānuku and Ranginui were in such a tight embrace that no light could penetrate. The two had several children, each with their own domains:

Tangaroa—guardian of the sea, whose children are all the creatures of the sea

Tāne Mahuta—guardian of the forest, whose children are the trees and animals in the forest

Tāwhirimātea—guardian of the winds, storms, and gales (anything pertaining to weather)

Tūmatauenga—guardian of war
Rongo—guardian of cultivated food
Haumia-tiketike—guardian of uncultivated food
Rūaumoko—guardian of volcanic activity and earthquakes. Whenever he gets uncomfortable and moves, earthquakes and volcanic eruptions occur.

The children decided that they wanted to separate Papatūānuku and Ranginui so that there would be light and space to move around. Each child had a go in trying to separate their parents.

Tūmatauenga, the guardian of war, wanted to kill Papatūānuku and Ranginui, but the others rejected this idea. Tāne Mahuta, the guardian of the forest, decided to lie on his shoulders and push up with his feet, which worked. The earth mother and the sky father were forced far apart. Tāne Mahuta clothed Papatūānuku with trees because she was naked. Tāwhirimātea threw stars and the moon at Ranginui to keep him company.

As we walked back up the steps, I turned back to take a photo of Wairēinga, half-expecting to see a guardian in the mist.

TUNNEL

In the middle of a long, hot ride through dairy farm country, I took a rest in a patch of shade outside of Whareorino School in Waikawau. It was the summer holidays; classes weren't yet in session.

The principal, Cynthia Kowalewski, saw me and came outside. She mentioned that she cycles herself. "Would you like to come to the pool out back?" she asked.

Cynthia had a big smile and wrinkles around her eyes from smiling in the sun. She has four kids. Her husband is a farmer. While we swam, her kids told me about their chickens and seventeen dogs.

Later, sitting on a bouncy ball in the classroom, I flipped through a picture book called *What is OIL?* This is the northernmost part of Taranaki, oil and natural gas country. The school has a wood-burning stove in the corner and a row of colorful ukuleles on the wall. A laminated sign illustrated the Māori numbers one through ten—*tahi, rua, toru, wha, rima, ono, whitu, waru, iwa, tekau*—and the days of the week. I tried each word out loud, noting the parallels with the Tuvaluan numbers, which I remembered hearing when Losite and her family played cards.

Cynthia pulled out a map to show her kids where Boston is. We talked about New York City. Cynthia had been there; she once nannied in Utah and traveled around America, "back before September 11th, to give you an idea of how old I am."

"She's thirty-seven!" one of her daughters called out.

"What happened on September 11th?" her eldest child asked.

"A terrorist attack on the Twin Towers."

"What's a terrorist?"

"Someone bombed the building," Cynthia said.

"To make them afraid?" one of her kids asked.

"Well, almost," I said. "They ran a plane into them."

"Are the towers still there?"

"One of them might be," Cynthia said.

"Both fell. It's a memorial now," I added. "Everyone in the States remembers where they were on September 11th."

Cynthia's oldest child told me about his trip to Disneyland in California. He brought his sisters back a Minnie Mouse T-shirt and a Mickey Mouse wristwatch. Then we all piled in Cynthia's car to go to Waikawau Beach.

Cynthia and I bonded talking about the hill just north of us in the forest reserve—how a steep hill can be just what you need to reset your brain on a long, solo ride.

The access point to the beach is a tunnel carved into stone and pointed at the top. It's two cows wide. Three men with picks and shovels carved it out in 1911, when farmers needed a way to bring their cows out to the ships and supplies in from the traders at sea.[7]

We made an audio recording under the sandstone cliffs, water dripping down the walls.

"It's big enough that you can get a four-wheel motorbike through here, but not a car. So it's a walking tunnel," Cynthia said. "It's a long way to the next access point," she added. "When you get to the end of the tunnel, you are greeted by the beautiful ocean and the crashing waves."

We played with the echoes. Cynthia called up a friend a few hours ride away, Merepaea, who agreed to host me for the night.

MATAKITE

Merepaea and Neville's house is set back from State Highway 3 in Mokau on a cliff overlooking the sea. I arrived late, having waited out a rainstorm at a gas station.

Neville is soft-spoken. In the morning, he shared a story.

"I had three kids, but only one of them survived," he told me. We were standing side by side on his back porch, looking out over the water. "One of them, at age three, fell off a chair and died. He used to run to the front of the house to meet me when I came home from work. His mother would run after him to catch him. He wasn't a steady walker. One day she didn't catch him in time."

A cliff dropped below us. I could smell the sea.

"My second child died at age seventeen," Neville continued. "He was coming home with his friend from a party in New Plymouth, and his friend drove straight into a brick wall. Drunk driving, it must have been. My son was sitting in the front seat."

"I'm so sorry," I said.

"My third son suffered after his brother died. But now he has a partner in New Plymouth who lost her brother the same way, drunk driving. They're both on medication for depression. They support each other."

Neville paused to take a sip of his coffee, a flat white. His shoulders hunched forward from the rest of his body. The midday light made his bald head shine, the same way it glistened off the pots and pans hanging from the wall. A laminated sign in the hallway read "Bald heads are beautiful. People with ugly heads have hair to cover them up!"

"I am a *matakite*, a healer," Neville said. "I see things. My great-grandaunt, Waikura, was also a *matakite*. I learned from her."

Water became, for Neville, a form of communication.

"A friend of the family went missing. His sister called me up to say that the family needed help. Well, I went to look. He was four kilometers out at sea. He talked to me.

'Where are you?' I asked.

'At the bottom of the sea,' he said.

'You need to come back for your family.'

'Okay,' he said.

"I told the family to be at the shore in four days' time, but he made it back in two. The family wasn't there to receive his body. Two tourists found him and alerted the police. The family buried his body a few days later.

"Later on I was sitting right there on that chair, watching TV," Neville continued, pointing to the spot, "and I see this figure with a hoodie on walking by. It was the boy. He said: 'Thank you for bringing me back to my family,' and then he disappeared. Merepaea was out working in the garden and I asked her if she saw anyone walk by. She said, 'No.' "

I paused, savoring the story. "How do you do it? The communication, I mean," I asked.

Neville waved his fingertips ever so slightly in front of his torso. "It's a focusing. Merepaea can do it, too. She can see a dark spot where the injury is. Well, I don't know if you believe in all of that. Do you?"

Later that night, just after sunset, I talked with Merepaea.

"What did you eat for lunch?" she asked.

"I had some toast. Neville and I shared stories over a flat white. He told me a bit about his healing work."

"Ah, yes. That."

"It's an art form, really. Being able to see so much."

"It's a gift, but it can get tiring," she said.

FLAG

I reached the base of Mount Messenger. A yellow-and-blue sign read: "Winding Road/Be Patient."

It was there for good reason. This section of State Highway 3 twists up a series of steep switchbacks. Summiting the mountain is the only way to cover the fifty-five miles from Awakino to New Plymouth along the coast. I was riding my bicycle to a hui of climate justice activists in New Plymouth, and I was running out of time.

After the first two switchbacks on Mount Messenger, I was, as the Kiwis say, "puffed"—out of breath. I squinted in the January summer sun and cursed every item I'd brought with me—will I ever use the water purifier? My harmonica? I trained my eyes on the road's precious few inches of shoulder and prayed that the truck drivers would see me in my neon vest.

At the summit, I stopped just before a narrow archway of a tunnel to enjoy the view of treetops and valley floor, the whole sky laid out above me like a thousand-piece puzzle. A cotton-ball cloud passed overhead. As I took long sips from my water bottle, a white van pulled up next to me, kicking gravel in its wake. The driver rolled down the window.

"I saw you starting the climb when I was on my way to a job," the man said, smiling through his beard. "Do you want a ride down? I'm a cyclist, too, and this bit is winding. No shoulders. Not so safe when the cars come flying."

A few minutes later my bicycle was balanced in the back of his van. "I'm Tony," he said. "I'm an upholsterer. What on earth are you doing at the top of this hill?"

I told Tony that I like going up hills more than I like going down them.

"That's ridiculous," he said. "You ought to go see the doctor for that."

When Tony asked if I had time to eat lunch with him and his family, I said yes. We pulled up to a farmhouse in the middle of a field. After sharing a few plates of avocado toast, cheese, and kiwifruit, he disappeared into his upholstery workshop for a half hour and emerged with a neon yellow flag he'd made for my bicycle, which he connected to my rear rack with a piece of old hose and a plastic dowel.

"This way the truck drivers can see you more easily," he said.

I made it to the hui on time.

HUI

For three days in late January 2015, around eighty people gathered at Muru Raupatu Marae, a communal meeting house in New Plymouth, to discuss the impacts of New Zealand's oil and gas industry on communities in Taranaki. The conference, titled "Taranaki's Beauty and the Beast" was co-organized by the Environment and Conservation Organisations of Aotearoa, Sustainable Whanganui Trust, and Climate Justice Taranaki.[8] I was at the pulse of environmental activism in Aotearoa.

After a *pōwhiri*, or ritual welcome, Diana Shand, an environmentalist from Christchurch, welcomed us with advice. "Accurate and powerful information will power our case," she said. "Working together is a powerful thing." We sat side by side on folding chairs, listening to PowerPoint presentations punctuated by conversations over communal meals.

According to the Petroleum Exploration and Production Association of New Zealand, over a thousand exploration wells have been drilled in the country, six hundred of which are in Taranaki.[9] This extraction began in 1865 with the Alpha well, dug near the Moturoa seeps in Taranaki.[10] The Moturoa oil field set the stage for a domestic petroleum industry in the early 1900s, which expanded in the 1950s with the adoption of seismic data.

Hydraulic fracturing, or fracking, is the process of using pressurized fluid to fracture rock. Water, sand, and chemicals are injected to crack open the rock and allow the previously inaccessible oil and gas to flow to the surface. The country's first documented frack was in 1989, at the Petroleum Corporation of

New Zealand's Kaimiro-2 gas well in Taranaki.[11] The majority of the fracking in Aotearoa has taken place in Taranaki.

Catherine Cheung, a researcher with Climate Justice Taranaki, referenced Dr. Anthony Ingraffea, a professor of engineering at Cornell, whose research estimates that more than 6 percent of wells leak, and virtually all of them leak over time. Wells fail. And the oil and gas industry doesn't stop. It operates 24/7. People living near the well sites are subject to light pollution and can't see the night sky. Hearing this, many attendees nodded in grim agreement.

Then Dr. Anne MacLennan, a palliative medicine specialist in Wellington and a member of OraTaiao, New Zealand's Climate and Health Council, took the microphone to give a presentation about the connections between fracking, health, and climate change.[12] Noise and light from the drilling sites keep both animals and human beings awake, she told us. Constant noise pushes up blood pressure and causes sleep deprivation, contributing to cognitive deficits in children and rising rates of heart attacks. The increased vehicle traffic and diesel-burning trucks, as well as tankers and heavy machinery, increase ambient air pollution. Children are more sensitive to particulate matter from diesel exhaust than adults, as they breathe faster and have a higher surface area in their lungs. Diesel exhaust is a class one carcinogen, she warned us. Water pollution is also a concern. As well head seals fail, groundwater and surface water near fracking sites can be contaminated. "With time, a lot of wells will start leaking, and we have thousands of wells which have been abandoned," she explained.

Companies that frack aren't required to disclose which chemicals they use. "If you don't know what the chemical is, you can't do a test. If you don't know it's there, you can't measure for it," Anne added. "Exposure tends to be episodic and fluctuating."

Some effects of exposure are immediate: choking, asthma, sore eyes, sore skin. "But there's also the sinister delayed onset, things that you don't know about for years. And the synergism between some of the chemicals," she told us, is a worry, "but again, we don't know what the chemicals are. It's cumulative, and some of the effects will be irreversible and life shortening."

She pointed to existing inequities in health and wealth that are exacerbated by this contamination. Children, the elderly, and Indigenous populations are particularly vulnerable. "Māori quality of life comes from close connection to land, and destroying that has huge implications for physical and mental health," she added. "Kids will inherit a world that has been badly damaged. What we're bequeathing is beyond belief."

Emitting methane from fracking accelerates climate change. And the health burdens of climate change are numerous: heat stress, extreme weather events, air pollution, food and water insecurity, infectious diseases, population displacement, economic collapse, conflict, and migration linked to climate change, which she reminded us, are not just features of some distant future. We are living with those impacts now.

"Children," she added, "are exposed to the changes longer. They may be left to cope with things that are un-copeable. That's not the next generation. It's the children who are already around."

That night, we slept side by side on mattresses in the body of the marae, the same room with sloped ceilings where we attended presentations during the day. I listened to the breath of the people around me. A baby woke up crying and his mother breastfed him back to sleep. I closed my eyes and tried to imagine many versions of the future. I am part of a generation of people who worries about the environmental impact of bringing children into this world. Will they have a life worth living?

In the morning, after a breakfast of muesli and yogurt, we boarded a bus to speed past cow fields and well sites. Taranaki, the volcano for which the region is named, winked in the distance.

Sarah Roberts, a schoolteacher who stood in the 2014 general elections as a Green Party candidate in New Plymouth, narrated the tour on the bus's staticky microphone. She noted that TAG Oil, a Canadian company, started drilling a fracking well four hundred meters from her home in 2011. Sarah told us that there is little in the way of regulation and monitoring at each well site. "They would visually inspect for contamination but not do any water sampling." Without those samples, the contamination was difficult to prove.

"In Taranaki, the money does not hit the ground here," she added. "It's remarkable that we have dairy and oil and gas and we are this poor." She pointed out each well site as we drove past. The fracking sites were punctuation marks on the landscape: easy to miss, but indelible.

Back under the roof of the marae, we reconvened to listen to Russell Gibbs, a Māori farmer who has worked in shearing, fencing, shepherding, and dairy farming. He works on fifth-generation *tupuna* (ancestral) lands south of the Tongaporutu River in Taranaki, and follows *tikanga*. To Russell, this means using no fertilizers and spraying only when necessary.

"It's about leaving the environment in a better way than it was on your watch or your generation," Russell told us. "If we have an excess of growth then we'll fallow," he said. "There's more life under the soil than on top. The effects of one fallow can last for years. You build a balance."

Russell's voice carried to the back of the room. "The negative side of climate change is that it's an abstract concept," he said. "But if there's climate change now, it was caused by actions twenty, thirty years ago." He noted that the weather patterns have changed little things, like the time of year when he cuts the bush. "You add up all those local changes around the country and then you have climate change in New Zealand." The damage we make now, he said, will compound every year. "If we don't try to protect our way of life, no one else will."

Next, Haumoana White, an elder in the community, spoke. As a shearer and slaughterman, he has farmed in Taranaki for most of his life.

"The Māori worldview of what climate change is all about is quite different to the commonly held view," Haumoana said. "For us, we think the word climate change is a misnomer that actually portrays something different. We prefer to call it 'the destruction.'"

To Haumoana, land and people are important above all else. "People cannot exist, anything cannot exist, if the environment is not right," he said.

"I think that if we all lived according to *tikanga*, we would make a difference in the environment around us. Our old people cultivated the grounds. I remember the huge gardens that they had. I had an old uncle who had one leg and every year he would dig, by spade, an acre of ground for his potatoes. They would leave ground fallow. We would give to Maru, a god of war. It's about putting back balance into our world," he said. "The petroleum industry imposes on us. We have two major pipelines going through our land."

Lately, Haumoana has expanded his business into honey. "We're finding that exciting and very close to nature," he said. "If you can understand bees, you can understand what goes on in your own community."

On the last night of the hui, we folded the chairs and tables and pushed them aside for a performance. After a weekend of sitting and listening, it was such a relief to move. The group danced to a live band, Te Kohikohinga Kohatu. One of the organizers, Urs, played the clarinet, his tall body lithe in the music. Three singers harmonized. A stand-up bass player slapped his instrument with gusto.

"This is an anti-drilling band!" Urs shouted at the end of one song. I spiraled, imagining the movement of water and gas three kilometers underground.

GEOLOGY

Ten months later I met Alex, a geologist, in Seacliff, a village a few hours northeast of Dunedin by bicycle. He was watering his garden. I stopped to ask if I could fill up my water bottles; they were almost empty.

Alex invited me into his living room. He told me that one of his earliest memories, at four years old, was digging a hole in his father's vegetable garden. When the hole got too deep for his shovel, he added a pole to the shovel handle to make it longer. After several days of digging, his parents surreptitiously added coins to the side of the hole, thinking he was digging for treasure.

"That's what kids are supposed to do," Alex said. "I was old enough to think, that's really sweet," he added, "but that's actually exactly not what I'm doing." Instead of digging for reward or treasure, Alex dug to satiate his curiosity, "to get really down deep and see what is inside Mother Earth."

Alex worked for forty years in the oil and gas industry, driven by a passion for geology and nature.

"I was able to do something that nowadays is regarded as a pariah of society," Alex said. "It wasn't like that in the early days, of course. Public attitude toward oil and gas was totally different to what it is now."

When we spoke in 2015, Alex had just retired. "I have to wear the mantle of being a fracker and loathed by many people because of it," he told me. "I was doing most of the fracking that's been done in New Zealand under my management."

Alex found it "absolutely amazingly puzzling" that fracking, and the rhetoric around it, "fills up so much passion and fury." In his retirement, "I'm left feeling a little bit sad that most things that we will, as people, complain about, are the things that we don't have much experience with ourselves."

For Alex, the most upsetting environmental issue is waste: excessive packaging, motor sport, the use of fuel for leisure, travel, and plastic bags. We're outstripping our resources.

"I sleep better having designed a frack in the earth than I do when I was given a bottle of cologne. I was furious at the incredible waste of packaging," he added. "Why would I want cologne in the first place? I mean, I don't stink."

The attitude of the industry, Alex told me, is "amazingly strict" about waste management and containing spills. "When things go wrong, it's awful," he said. But having been a part of the engineering side, he feels confident in their diligence. "It's all engineered. And if it went out of control, if the frack went and started going up toward the surface," he said, "we'd instantly see it because of the pressure response and behavior that we're monitoring when the frack is being pumped. We're not fracking in the dark, as it were. We're fracking in a very controlled and engineered way."

Most of the gas in Taranaki is converted to electricity or sold as liquefied petroleum gas, both in New Zealand and abroad. "It's principally for keeping homes warm and powering things," Alex explained. He noted that homes, lighting, factories, timber-processing, and milk-powder plants all use a lot of electricity generated from gas. In New Zealand's North Island, there are pipelines that deliver the gas to power stations and homes. In the South Island, there aren't gas pipelines, but liquefied gas is delivered in tanks for cooking and heating.

I asked Alex about his thoughts on climate change.

"I'm really angry at the arguments put forward by the climate change deniers. I find them trite and trivial. I'm a scientist," he said. "I believe global warming is inevitable. We're just an organism," Alex said. "I don't think we as an organism can forestall anything. We never have. And it will play out."

A fly buzzed past us. I registered the movement of its wings in my headphones.

"Nature will have its way with us," he mused.

EELS

One person from the hui recommended that I meet Dr. Mike Joy, a freshwater ecologist who was then working as a senior lecturer at Massey University. Mike is critical of the impacts of intensified dairy farming on New Zealand's rivers and eels. We spoke in his kitchen, where painted-on eels swam stagnant on the floor.

"Water and climate change are totally interrelated here, as I guess they are everywhere," he began. "As a country we are going along a very clear path toward maximizing agricultural production. Our government has decided that it wants to double agricultural production in the next decade or two. We already have extreme effects from the intensification of farming," Mike said. He estimates that 62 percent of the length of New Zealand's rivers are unswimmable because of pathogens, mostly from farming but also from urban impacts. "Pretty much anywhere in lowland New Zealand you'll find polluted waterways, and in the conservation estate, which is mostly the alpine areas, we have amazingly clean lakes and rivers."

This creates an irony. New Zealand trades on its clean green image in tourism marketing (and *Lord of the Rings* sets), but the reality is different. Many alpine estates are green, but runoff of nitrates from the dairy industry have made many lowland rivers unswimmable.

With the intensification of dairy agriculture, "we now have 6.5 million dairy cows in this country," Mike explained. "If you want to think about the impact of 6.5 million dairy cows, a very conservative comparison is that one dairy cow produces as much waste as fourteen humans. So you multiply that out," he said, doing the mental math, "a country the size of the United Kingdom and has ninety million human equivalents."

And the waste from cows doesn't get treated, like human waste would. "It mostly just goes into waterways," Mike said. This overload of nutrients asphyxiates ecosystems, causes toxic algae blooms, and makes it unsafe for people to swim in the water.[13]

Nearly 50 percent of New Zealand's greenhouse gas emissions come from agriculture.[14] "While our CO_2 emissions are relatively okay because we're quite a low human population, our greenhouse gas emissions in total are massively increasing," he pointed out. The country's human population, as of September 2020, is estimated to be just over five million.[15]

"We trade on this clean green image, but in reality we've trashed an amazingly unique and diverse country in the name of producing a very low-value commodity—that's milk powder." In 2016, New Zealand was the largest exporter of dairy in the world, accounting for 28 percent of world trade. "We're a major player in global markets, but only at the cheapest possible end of the scale," Mike said.

And at what cost? "The other reality that's well hidden here," Mike told me, is that New Zealand has the highest proportion globally of threatened indigenous species.[16] Around three-quarters of native freshwater fish in New Zealand are at risk of extinction, and these are just the species that scientists know about. Shortjaw *kōkopu*, giant *kōkopu*, *kōaro*, *īnanga* (all whitebait species), *kanakana/piharau* (lamprey), and *tuna* (longfin eel) are all at risk.[17]

Mike recommended that I stop a few kilometers down the road at The Quarter Acre Cafe in Manakau, where eels writhed through a creek in the backyard. I ordered a coffee and sipped it by the water. The eels moved smoothly beneath me, hoping to be fed.

TE RIPO KOKOHUIA

Tanea Tangaroa, a Māori woman in Whanganui, has spent nearly two decades restoring Kokohuia, a wetland reserve that had previously been a landfill site.[18] During that time she has hauled out tires, led teams to remove alligator weed, and planted native trees.[19] The toxicity is still beneath the surface, but some birds are returning. To Tanea, wetlands are lungs—a living classroom.

We walked together to the entrance of Te Ripo Kokohuia, in Whanganui's southwest, near where the river meets the Tasman Sea. The land felt porous and soft under my shoes. Tanea wore gumboots.

As we entered the land, Tanea sang in Māori, her vowels lilting upward, then down. I recognized one word: *wai*. Water. We walked through a gate.

"*Kia ora*," she said. "So a translation to all of that. Wow. You probably felt it in there anyway. It's an acknowledgment of those things you don't see—the things that we should see that we don't."

We paused together to breathe and to listen.

"Every bird and creature has the same language," Tanea told me. "When you observe and watch them, they tell you things."

Before the Resource Management Act was passed in 1991, much of New Zealand's waste was disposed of without controls.[20] While modern landfills are lined, the Balgownie Tip has no lining at the bottom; layers of waste were compacted into the dump, which was built after the Second World War.[21] Though it ceased to be a landfill in 2000, and was capped with clay and topsoil in the early 2000s, Tanea still worries about asbestos and leachate that can impact the water, birds, and plants that make this place their home.[22]

"As you can imagine," she said, the area is "like one big toxic soup. So you can't eat anything from out of here at all."

The *repo* is not the thriving wetland it once was. Tanea set traps for rats and stoats and ferrets, all introduced species that decimate Aotearoa's population of native birds. But still, there are signs of life. Tanea told me about the owls, hawks, kingfishers, and cormorants that started to return as the *repo* recovered.

"We don't like calling this a landfill because it demeans what it really is," Tanea told me. But the signs of the land's former use are all around. "In those three stands of bush are three methane vents which come right out of the hill and you can hear it," she told me, mimicking the *pshhhh* of the gas escaping, "and that's all the methane that's coming up from the landfill."

We stepped through the thicket, past brown water and a plant that Tanea identified as purple hebe. "*Rongoā* or medicine from here is in the tip," she said, pointing inside the flower. "And now you chew that, it's very bitter." Tanea described its use for diarrhea or vomiting. "This *repo*, not only does it have a lot of birds," she told me, "but it's also a medicine cabinet."

Tanea hopes that in the future, a team of volunteers will reclaim the wetland. "You see all the birds coming in. You see them moving. They're talking and the wind blows. That's the song of the *repo*," she said. "It's a sanctuary," she added—not only for native birds, but also for people. "We need to come and have time of prayer or spiritual cleansing." The waterways have meaning beyond any individual species. "They're just not food cupboards. They're just not places to learn. They are also churches. They're also places where you come and where you can cry or you can laugh or you can talk to your loved ones and sing for yourself," Tanea continued. "The *repo* here, it's beautiful."

ICE CORES

When we met in 2015, Peter Neff, a glaciologist and earth scientist, was completing his PhD in geology at the Victoria University of Wellington's Antarctic Research Centre. I wandered onto campus and happened to catch him in the graduate student lounge before one of his classes. He told me that Antarctica is a massive place—bigger than the United States. New Zealand's sliver of the continent, the Ross Dependency, is approximately the size of Sweden.

Peter does his research on the central plateau of West Antarctica. "You're getting nosebleeds because it's so dry," he told me. "You get in your sleeping bag and it's just all static electricity."

Roughly 70 percent of the world's fresh water is locked up in Antarctica.[23]

If all the world's glaciers and ice sheets were to melt completely, scientists estimate that global sea levels could rise by as much as 195 feet.[24] This melt could happen "in a geological blink of an eye," Peter said, snapping his fingers.

The challenge for Antarctic scientists is to sort out how many of these changes are natural, and how many are driven by humans. To answer this question, Peter and his colleagues look backward and drill cores in the ice—about five inches wide and some more than two miles deep. Ice cores, retrieved in approximately ten-foot sections, can provide detailed information about how water in Antarctica was distributed in the past thousands of years. In the lab, Peter and his colleagues melt the samples and analyze their component parts. This lets them understand how ice shelves, snow accumulation, and snowmelt are changing in different parts of Antarctica on a year-to-year basis.

With this data, Peter and his colleagues can piece together "spatial differences in the timing of the behavior of the ice sheet"—in other words, how certain areas in Antarctica respond to a warming planet.

"The science of ice cores, which is just what I work on, is only sixty years old or so. We've gone from having an ice core that represents all of Antarctica—because it's the only one we have—to now we have ten or fifteen deep ice cores from Antarctica from most parts of the continent," he explained.

The West Antarctic Ice Sheet—located south of the Pacific Ocean between South America and New Zealand—is one cause for concern. It's "a dynamic bit of ice," Peter said. With too much warming the ice shelves that keep West Antarctica in place can collapse.

"There's no way you can explain the warming trends that we see besides input of anthropogenic CO_2 and other fossil fuels," Peter told me.

"I'm not just doing this as a job. It's very inadvisable if you just want to make money," he said. "You have to do it because it's really fun."

THERMOHALINE CIRCULATION

Density drives the motion. As water goes toward the poles, it gets colder, which makes it denser. As sea ice forms, fresh water is removed. Cold, briny water sinks deep into the ocean. This sets up a tug of war—as that water is pulled down into the ocean, more warm water can travel to the poles. This process, known as thermohaline circulation, maintains features of the global climate that we have come to expect as stable, like the Gulf Stream, which travels across the Atlantic and keeps Europe warm.

Christina Riesselman, a paleoceanographer at University of Otago in Dunedin, reconstructs how ocean circulation was different in the earth's past. To find an atmospheric carbon dioxide concentration of four hundred parts per million—the threshold we passed in May 2013—you have to look back three million years into the past.[25] At that level of carbon dioxide, the planet was on average 2 degrees Celsius warmer. The Intergovernmental Panel on Climate Change has pinned 2 degrees as a safe level of change.

Looking back three million years gives us an analogous climate to what we might see in the coming century. "It turns out that 2 degrees of average warmth was really unevenly distributed on the planet and the poles were considerably warmer," Christina told me. This has implications for the West Antarctic Ice Sheet; its collapse could be sooner than scientists had previously anticipated.

Christina also described the warm anomaly—the difference between the modern sea surface temperatures and sea surface temperatures three million years ago—as a "really alarming facet of that past warmer world." In an analogous climate, the North Atlantic was about 8 degrees warmer, on average, than it is now. Eight degrees "was enough to cause a significant perturbation to the way that the Atlantic Ocean flowed," Christina said.

On that three-million-year analogue planet, the North Atlantic produced much less vigorous deep water, which means that circulation was greatly reduced. "The Antarctic sort of stepped in to fill the void," she said—accelerating ice loss even further.

In the future, Christina pointed out, the climate could evolve such that Europe could be considerably colder. And this, of course, has impacts for both water and agriculture.

"As a species we have evolved our population centers just within a snapshot of climate history," Christina said. "We're not very flexible."

COWS

I looked forward to Inchbonnie, a town on the west coast of the South Island, my entire trip. I annotated the name in my roadmap with a question mark. Inchbonnie had a poetic ring to it. I was intrigued.

People warned me: there's nothing there. A patchwork of farms. Still, I was determined. When I arrived, pushing through on my two tires of air, I didn't see a single person. No one told me a story.

The cows did, though. Hundreds of them, packed onto grass, green and fertile. The nitrates of their urine polluted the water. The methane of their breath warmed the air.

And the sign denoting the town, Inchbonnie, had a gunshot through the last *i*.

RETREAT

Ian Dalzell is a potter who lives and works in Moana, on the west coast of New Zealand's South Island. He fires his work with coal.

I turned into Ian's driveway on a whim, following a red refrigerator full of books: a little free library, waterproof.

When I arrived on my bicycle, Ian and his wife, Sue, offered to let me stay for the night. They gifted me a bar of Manuka soap and showed me where I could bathe in the river unbothered. I washed off the salt glaze of the day, and then spent the evening throwing a stick for their dog, Dotty, over and over again. After dinner, Ian pulled out a photo album from 1968.

"We have two glaciers that are about a hundred kilometers below us on the main divide in the South Island," Ian told me. "One of them is the Franz Josef and the other one is the Fox. They both come from a similar source."

Ian first visited the glaciers in 1968 when he was eleven or twelve years old. "I can remember coming around the corner of the road and there was just a mountain of ice in front of us," he said, gesturing with his hands to communicate the enormity of it. "We walked up to the glacier, and it was quite easy to walk. It was only a few-minutes walk to the terminal face of the glacier."

In the years since, Ian has returned to the glaciers a handful of times. "We were down there only about six months ago, and we came around the same corner and the glacier has receded right up around so that it's actually out of sight," he told me. "You can't actually see any ice from where we originally saw the glacier."

There were a few years in the '90s, Ian remembers, where the glacier stopped receding and advanced down the mountain. "They thought that it might have been just the result of heavy winter snowfalls," he said. "It's been retreating ever since."

GLACIAL LAKE

Lured by the mountains, I took a detour to Aoraki/Mount Cook National Park. Jono Hall worked at the Youth Hostels Association in Mt. Cook Village. We met outside the national park.

"The Tasman Glacier is retreating," he told me, "and it's kind of undeniable that that is down to climate change and global warming." Since the 1990s, the New Zealand National Institute of Water and Atmospheric Research has documented an average retreat of 180 meters per year.[26]

Now a glacial lake, seven kilometers long, has pooled at the glacier's terminus.[27] Jono told me that his girlfriend's dad mountaineered near the Tasman Glacier in the 1970s. At that time, the glacial lake wasn't there. "He came back this year, and he was like: 'What the hell is that?' And that is from global warming, you know," Jono said. "There's no disputing that."

The Franz Josef glacier has also retreated "from the valley floor way up the mountainside," he explained. "We're losing a huge amount of snow off the mountains, and all the hanging glaciers are retreating as well. They're not as affected as the lower-down places, but you can see it here probably more than anywhere else in New Zealand because of the size of the glaciers."

Up close, the glaciers have a voice of their own. "You hear them creaking and groaning, that's natural. But the amount that's falling off the face of the glacier every year is accelerating," Jono said, "and it's pretty sad. It's pretty scary. The landscape is disappearing. Who knows what's going to happen?"

FLOOD

When we spoke in Omarama in 2015, Noelline Gillies was eighty years old. "It's taken nearly all my life," she told me, "to finally sort out why I'm frightened of water. I love water. I love the rivers. I love the sea, but I am always very concerned if I have to get into the water."

Only recently did she solve the mystery of her fear, with a memory that resurfaced from when she was five years old.

"I was with my grandparents in a cottage which was low on the ground, and just across the road was this big river in flood, the Taieri River, near Dunedin, New Zealand," she said. "And the floodwaters were getting higher." Everything that could be lifted had to be stacked up on beds to keep it dry.

"And then fortunately for us a man came along with a dinghy, and he came right up to the doorstep and he was there to pick us up. Were we pleased to see him," she said.

"My mother had my sister in her arms. She was only a baby. It was pouring rain. My grandparents, Grandmother and Grandfather, were elderly, and we didn't even take anything with us because there wasn't any room in the boat," she added.

The current was so strong that it almost sucked the boat under.

"This man, you could see, he strained as hard as he could with these oars, because there was no motor, to fight the river. And we sat there in silence because we could see that if we were dragged into the current of the river, that would be the end of us. So he had to work very hard, and he actually managed to get us past this danger."

The man rowed to a safe place on a hill where Noelline and her family could disembark. Rivulets of "dirty, clay water" ran down the hillside. Noelline recalled walking several miles to safety at a farmhouse farther uphill.

In the meantime, the man rowed to the railway station for another rescue. The water was so deep that the stationmaster was sitting on the roof of the two-story building. The man in the rowboat "was a hero," Noelline said, "though he probably never got recognition for being a hero."

Though it happened three-quarters of a century ago, the memory of the flood is still crisp. "In my mind, when I really think about it, I can see this dirty, filthy, horrible water and the whirlpools with the water swirling around and just the fear we all had that we might get sucked into it," Noelline concluded.

"And it's taken half my life to understand why I was a very reluctant person to learn to swim."

VALUE

A few days later, I met Tracy Hicks, the mayor of Gore District Council, inside his office. "We've spent the best part of one hundred years draining this province to get the best production we can out of the land. We've got some really good, high-producing, fertile land here now," he told me. "The changing landscape here and change of land use from reasonably extensive sheep and beef farming to very intensive dairying has changed the face of this province completely over the last ten years." This has created water quality and quantity challenges for both rural communities and the urban areas they are connected to.

The Gore township, Tracy told me, sources most of its water from wells that are connected to the Mataura River. When the Mataura River goes down, then the township has to restrict the amount of water that it takes out of the wells.

"What's causing the river to come down to that level is the question that everybody's asking. Is it the extra pressure that's being put on the land production-wise? Is there a changing climate that's causing that? What is the challenge? But, whatever the cause is, the result is that for probably four months of the year, the urban community has to restrict its water use because we simply can't get it out of the ground," he explained.

"I think that we've taken for granted that water is going to be there in the volumes we want, when we want it, forever," the mayor added. "And it's made people much more aware of the value of water."

"Are there any kind of personal stories or experiences that you've had with water that come to mind from this community?" I asked.

Tracy took a moment to think.

"I was born and raised in a little town just south of Gore—Mataura—which is named after the river," he told me. "Mataura is a community that grew up around the banks of the river, on a set of falls in the river, and the river was used to generate hydroelectric power for a paper mill and freezing works."

In addition to power generation, the river was used as a waste disposal channel. Tracy told me that his parents owned a shop in town. One of his after-school jobs was to empty the rubbish from the store. "And like everybody else in town, the rubbish went over the bridge, into the river," he noted.

"I can't even conceive of anybody thinking about doing something like that these days. That's probably only fifty years ago that that was happening, which I suppose is, well, to me, it seems like a short time, but there's been a huge change in attitude to the value of water—to the value of streams and rivers and the fish and the inhabitants of the river—in that time."

The river, he added, has "still got a long way to go in terms of quality, but that's a heck of a lot better than it was."

CHANGE

When we met in Southland at the Waikawa Museum, Allan Stronach was seventy-seven years old. He worked as a commercial fisherman. "I have lived here all my life," he told me. If anyone was an authority on local weather—or the embodied experience of it—it would be this man. We spoke next to a display about farm machinery, with museum placards explaining the industries of whaling and sawmilling nearby.

"With rivers, I've found them all much the same as they were years ago," he said, before adding that the weather patterns had changed. "Fifty odd years ago it was more settled," Allan continued.

"You'd get quite long fine spells." In the last few years, he told me, "you'd be liable to get three or four weeks of rough weather and a few fine days," or "a few weeks of fine weather and just two or three rough days." The weather patterns, in other words, are more unpredictable than Allan remembers earlier in his life.

"It's completely changed around," he said.

BITTER

Cold has many flavors. There is cold that freezes the inside of your nose. There is cold that causes your toes to swell with chilblains. Cold that necessitates the thickest wool hat. Cold that a farmer will curse.

Carl Portergeiss, a sheep farmer outside Queenstown, told me that Central Otago used to be known for bitter winters and dry summers.

"But that's all changing," he said, "which is kind of good from a farming perspective because it's milder."

Back in the 1920s, Carl told me, the land where we were standing was so cold in the winter that trees would die. Sheep stations, in winters with high snowfalls, could lose half of their stock.

"You don't get any of that anymore," he said. "The most I've ever seen here is about six inches and it never lasts more than a couple of days."

Carl qualified his observations, saying that he couldn't be sure if the changes in Central Otago's weather were a direct result of climate change, or just a fluctuation.

"Not all places on the planet are going to suffer, at least in the short term," he admitted. "But then I guess if you're producing something for the export market and everything else is stuffed, it's not really going to help you."

ICEBERG

I wandered into First Church of Otago in Dunedin—a fortress of stone on top of a hill with a sharp steeple that slices the sky.

Malcolm, who worked in a museum in the church devoted to the history of the congregation, told me that in 2006, an iceberg from Antarctica floated past Dunedin's coast. The pieces likely broke off from an ice shelf in 2000.[28]

"It could be seen by people from Dunedin if they climbed up the hills and looked out to sea," he said. It was white and bigger than a speck, but far enough off the coast that it didn't come ashore.

This ice was a whisper from Antarctica—the faraway, suddenly nearby and in motion. Melting.

"Many people chartered airplanes to fly out over it and look at it," Malcolm said. The news website *Stuff* reported that a Dunedin helicopter company harvested pieces of the iceberg for vodka cocktails.[29] One sheep farmer brought a sheep up on the ice to film a dramatic shot of shearing.[30] Malcolm pointed to a photograph taken by the *Otago Daily Times* in which a helicopter, insect-size in comparison, landed on the surface of the ice. "You can see it's quite a huge thing." The iceberg floated off the coast of Dunedin for several weeks, melting as it drifted north.

KAUPAPA

I spoke with Jenny Campbell, a septuagenarian from Mossburn, Southland, at a café in Wanaka. We had traveled together to listen to her grandson, Corwin Newall, play piano as part of a symphony. "My biggest water story at the moment is shedding tears when I heard him playing," she said.

As a child, Jenny lived on a farm. She swam and drank water from the river without worrying about pollution. Her uncle had a dairy farm next door, and her family raised sheep. "There was no pollution of those waterways, as far as I remember," she said, "and they didn't have thousands of cows."

Now, "the whole emphasis on dollars as opposed to our environment" worries her. Jenny became involved with Coal Action Network Aotearoa's campaign to stop Solid Energy, a coal mining company in New Zealand, from building the Mataura briquette plant about an hour from her home, on the banks of the Mataura River.[31] The plant would convert lignite into briquettes that could be burned for fuel. "Their whole scheme was to build this big factory and dig the lignite up from under this wonderfully productive soil and turn it into little briquettes which would then be burnt in peoples' fires locally and exported overseas," she said.

"It's not just about the climate change but it's also about water. Water is such a basic requirement for us to be alive, along with every other living creature," Jenny added. "And we have a responsibility as people to protect the water—globally and particularly in our own backyard."

Kaupapa is a Māori word that refers to a set of values, principles, and ideas that serve as a base for action. "As a committed Christian, this is one of our kaupapa, our understanding of what we're here for is to care for creation," she said.

In January 2012, Jenny joined a group of 120 activists who camped on Mike Dunbar's farm in the Mataura Valley. Mike had refused to sell his family's farm to Solid Energy—the company's plan was to dig out the lignite beneath it. He was one of the last farmers who held out. At the summer festival, organized by Coal Action Network Aotearoa, Jenny started to think critically about New Zealand's greenhouse gas emissions and the impact of mining on aquifers.[32]

"We said that we stopped the briquette plant, but actually it was world prices going down," Jenny said. "So that was really quite funny that they put twenty-three million into building this plant and now it's not being used at all. It's got the shiniest chimney of any factory in the world, and it's still sitting there in pristine condition."[33]

ELEPHANT

After reading my cardboard sign, Jennifer approached me on the steps of Te Papa Tongarewa, the Museum of New Zealand in Wellington. She was eight years old and visiting from Porirua with her parents.

"This is my story about water," she told me.

"Once upon a time, there was heaps of water, like at the beach. One day it got so hot that all the water drained out. And then, the next day, the water came back. So, then this big elephant came and he drank all the water, and then he took it to this different land that no one knows," she said.

"And after all the water's gone out of the beach, everyone was getting really hot. And everyone walked around and thought there was a desert." Here she paused to look at all the museum-goers surrounding us, moving about their lives.

"Then one day it rained so hard that all the water came back, and everyone was happy."

4
AUSTRALIA

FIRE

In the two weeks before February 7, 2009, a heat wave scorched Victoria.[1] Temperatures in Melbourne topped 109°F for three consecutive days. That Saturday, the temperature reached nearly 115°. A fallen power line started a blaze in a field, which spread through a pine plantation. Winds of over a hundred kilometers per hour swept the flames through unpredictable fire paths.

That day, Dale Cox's neighborhood in Eltham was filled with smoke. There was little time to evacuate—the flames were moving too fast. A combination of high temperatures, strong winds, and dry fuel made the fire uncontrollable. On Black Saturday, 173 people died. Over two thousand homes were destroyed, and 1,660 square miles burned.

"If the prevailing winds hadn't changed at about four thirty when they did, the ground that I'm standing on now would have been reduced to ashes, along with a good portion of the eastern suburbs of Melbourne," Dale told me. He lost five friends that day. To deal with that pain, he paints. A series of his canvases depict two skeletons in suits, shaking hands.[2] Their suits are made of brown earth speckled with trees. The sleeves are on fire. Smoke curls off of their shoulders.

"The fire was heading here with such ferocity that there would have been absolutely no way that we could have stopped it. They're the sort of fires that generate their own climate, so you end up with absolutely nothing that can make a difference, no water-bombing aircraft could get anywhere near them," Dale added.

Climate change, which caused more days of extreme heat and more frequent droughts, exacerbated the fires and made them more deadly. "It's not a natural phenomenon, my friends, it's climate change. End of story."

DROUGHT

I met Constance at an event at Montsalvat, an artists' colony in Eltham, Victoria. She pulled me aside to speak about drought and death.

"I come from a tiny little town of about three hundred people in northeastern Victoria called Eldorado," Constance said, "and there's a creek called Reedy Creek that runs through the entire town."

One summer, in the middle of an almost fifteen-year drought—the "Millennium Drought" in Victoria that lasted from 1996 to 2010—Constance walked a few kilometers downstream collecting fish bones, turtle bones, and turtle shells.[3] The riverbed was bone dry. She kept the bones and shells in a basket in her mom's shed.

"She hates them," Constance said, "but I was really struck with just how many dead things I found in an environment that just wasn't meant to be dead."

Australia is a landscape of extremes.

VIVA LA VIDA

The running commentary went something like this: some guy pointed at my bicycle. "What is this, Tour de France rejects?"

Yes, clearly.

Or, as I rode by, kids would ask: "Why are there so many bags on that bicycle?!"

I never stopped to explain myself.

From May to October 2015, I biked two thousand miles up the east coast of Australia, from Melbourne to Townsville, and added ninety audio interviews to my archive in progress.

At the beginning of this journey, in Victoria, I cycled for a few weeks with

Isabel Vermeulen, a science teacher from Belgium. We found each other through Facebook. Some of my contacts—people I had never met but somehow knew through the magic of the internet—insisted that we connect. The community of women who choose to pedal alone is, while not tiny, smaller than you might think.

Isabel was a more confident stealth camper than I was. At the end of the day, she would rock up to a small town and pull into a pub for a beer and a plate of french fries. While sipping, she used Google Maps on satellite view to find a park nearby with tree cover and, ideally, a toilet block. After dark, she set up camp. Before dawn, she packed up to leave with the first rays of light. I was nervous at first to follow her, but no one bothered us; I became bolder in proximity to her swagger. Isabel also taught me to pedal going downhill—previously, I had just coasted. With a loaded bicycle, I am slow. She was always faster.

Weeks later, north of Sydney, I was pedaling alone up a gradual incline when two guys decked out head to toe in pink passed me on a tandem. They dreamed up the idea to cycle together and raise money, they told me, during a night of heavy drinking. Originally, it was a dare. Then they started to take it seriously. They bought a used tandem bike on Gumtree, the Australian equivalent of Craigslist. There was only one thing left to choose: a cause.

"We settled on something we both loved," they told me. "Breasts."

Their pink kit was from Australia's National Breast Cancer Foundation. Each night at a pub, they raised hundreds of dollars. The strategy: pull in somewhere, wear the pink, drink, and chat up the crowd. Then they woke up at dawn to do the same thing over again.

Each day I chased them—they texted me their location—and by the time I caught up they would be tipsy, their helmets full of cash. They let me sleep on the floor of the motel rooms they rented. At the end of their ten-day trip from Sydney to Byron Bay, they got matching ankle tattoos of tandem bicycles to signify their matehood.

Some people asked if I was lonely. I wasn't. It might sound strange, but traveling alone is the best way to be social—it's easier to approach people.

Most of the time, because of the distance = rate x time dimension of motion, I met cyclists moving in the opposite direction. Sometimes, we would pull over to talk briefly, sharing tips about the route and campsites and where to find the best plate of potato wedges.

Once, in Queensland, I pulled into a tiny town called Mount Larcom just before dark, unsure of where I would pitch my tent for the night. I was on a road that felt like it would never end, lined with unhoppable fences that made stopping to camp impossible. I chased the pink sunset and asked the clouds to please let there be a good surprise around the next corner.

And suddenly, there was. My surprise was sitting with a loaded bicycle at a picnic table: Nico. The twenty-nine-year-old Italian had just finished a coast-to-coast ride through the center of Australia. He was cycling south in search of work to replenish his funds. Every twenty kilometers, he told me, he stopped to smoke a cigarette.

Nico didn't have panniers on his bicycle and instead towed everything in a trailer made to carry a child. We quickly realized that switching to Spanish was easier than English and decided to stealth camp for the night in a rugby field. A light rain fell. A whole family of kangaroos grazed and hopped at the field's opposite edge. We exchanged stories from the road and cooked spaghetti.

In the morning, Nico and I parted ways. We left notes on each others' bicycles. He wrote on my top tube in Sharpie: "*Viva la vida.*" The words matched a tattoo on his forearm.

One of the things I like most about bike touring is that I have an excuse to be outside all day, every day. The little things become so beautiful. I notice the patterns of birds overhead. Friendships, while fleeting, feel meaningful. There is no windshield. Just immersion.

BLOCKADE

From 2008 to 2013, Dylan Grimwood, a climate activist, sat in trees in southwest Tasmania to protest the logging of some of the last remaining old-growth forests on the planet. For months at a time, Dylan lived in a platform affixed to a tree forty meters above the ground as part of the Camp Florentine blockade in the Upper Florentine Valley.[4]

The main threat at the time was Gunns, a timber company that was once the biggest hardwood sawmiller in the Southern Hemisphere.[5] Forestry Tasmania, a government organization, was also felling large segments in the name of forest management.

In addition to living on tree platforms, other tactics at Camp Florentine included digging tunnels under the road where people could sit to stop heavy machinery from going over. Sometimes, the activists locked themselves to bulldozers. Large groups of protesters, linked elbow to elbow, would block the roads with banners reading "Save the Divine Upper Florentine" and "Climate Action Now."[6]

The age of these forests make them valuable in a changing climate. Old trees store more carbon than younger ones. "These are ancient forests we're talking about," Dylan told me. They're massive carbon sinks—among the most valuable left intact in the world.[7]

Tasmania "was a little bit out of the way for the colonizers," Dylan said, so tracts of old growth forests in the south and northwest parts of the island state "survived the early years of colonization without getting logged."

Dylan described the experience of blockading as intense. "It's something that really changed me more than I could have known." He started tree sitting in his early twenties, enthusiastic and energetic about the activism around him. "I went into it pretty naive," he said. "I just wanted to apply myself to something that I felt was genuinely positive."

But going out of his way to do the best possible thing, Dylan told me, left him open to the worst possible treatment. After a few years, he realized that "The more we got in trouble, the better we were doing. Or the more we pissed the authorities off, that was a gauge of figuring out how successful we were being."

Living in a rainforest, Dylan got used to being constantly wet. He learned how to build a fire out of damp wood. Simple things like a hot meal at the end of a cold day became holy.

In January 2009, fifty police officers and a search and rescue team came to break up Camp Florentine. Dylan sat with Miranda Gibson, another environmental activist, in the branches of a sixty-meter-tall eucalyptus tree at the end of the road.[8]

"They just bulldozed around us," Dylan recalled. Trees thirty to forty meters away from them began to fall. "If it went the wrong way, we would have died," he said. At one point, a bulldozer drove right at him.

"I definitely didn't like the cops before I blockaded, but through blockading, I had a lot of life experiences that showed me what they're capable of," he said. "They're definitely not a trustworthy group of people."

There was beauty in living in a tree sit. One tree, at the front of the blockade, was full of black and yellow cockatoos. "They don't sound that graceful," Dylan said, throwing his head back to imitate their sound. "They're like *craaaa craaaaq*."

One night, Dylan woke up to the cockatoos, mildly annoyed. He looked above him to see a bird at the tiniest part of the top of the tree—balancing on a branch no wider than his thumb and forefinger.

The cockatoo flapped its wings and talked to its mate in an adjacent tree. A few seconds later, the bird went into free fall for a moment before spreading its wings and gliding off into the wilderness—over the canopy of myrtles and celery-topped pines and sassafras trees, the understory of mosses and ferns and fungi.

"The cool thing about being in the rainforest is that wildlife are just doing what they do and they don't always notice that you're there," Dylan said.

Direct action (and a collapse in native timber markets) garnered results—in 2013, a hundred and twenty thousand hectares of old growth forest in Tasmania was formally included in an expansion of the existing Tasmanian Wilderness World Heritage Area, giving the trees around Camp Florentine the highest level of conservation.[9]

LEADBEATER'S POSSUM

I sat alongside Don Butcher, an ecologist and outdoor educator, on a porch looking over the backyard of his childhood home in Eastwood, a suburb of

Sydney. Trees poked out above us, stark against the morning sky. Don unfolded a map of Melbourne to show me the Yarra River catchment, tracing his finger down the river's spine.

"The Yarra, when it flows down at Melbourne, looks like a muddy neglected stream," he told me. "People who live in Melbourne would probably think, *you wouldn't drink that.* But it couldn't be further from the truth."

He pointed to Mount Monda, just over nine hundred meters tall, noting that Monda comes from the Woiwurrung word for rain. "Mount Monda is Melbourne's rain mountain," he explained. Its forested slopes capture around two meters of rainfall annually.

In addition to supplying water for the people living downstream, older forests are more biodiverse and better able to cope with stresses like drought and wildfire.

"We have predicted more catastrophic weather, so more heat waves and more big fires, but that aside, I feel we need to be good stewards of this landscape and we need to grow the forest into older age," he said.

In 1939, the Black Friday fires swept through Victoria, killing thirty-six people in a single day; as a result, much of the oldest forest is less than a hundred years old.

"We need to grow them for another seventy years. And if we do that, the forest will be like a golden goose. It will give us more water," Don added, because young, growing forests are thirsty. As a forest matures, the water yield increases, which would translate to more water for Melbourne and more water for the Yarra River.

And this is about more than just people. Wetlands are a source of all kinds of abundance: birds, fish, and eels all follow the water. "The Yarra once had wetlands," Don said. These days that water is diverted long before it becomes a wetland.

Don told me about the Leadbeater's possum, not "those bloody things in the roof," but smaller—"like a little caped crusader. I like to think of them as a ninja possum," he told me, "because they're very quick." The Leadbeater's possum is a symbol of old growth. It lives in the ash forests and subalpine woodlands of Victoria's central highlands. A small lowland population of the possums calls the land east of Melbourne home, but it is dwindling.[10]

Don spent years looking for them. Some people call them fairy possums. "I actually thought I was looking for fairies. I just figured they didn't exist, but there is magic in our forest," he said. "They do exist. I have seen them."

Don worried about what he calls a "profound disconnection"—that the majority of people living in cities don't even know where their water comes from. "People drink water. They shower in it. They bathe in it. They water their garden and they have no idea where that water comes from," he told me.

"We need to connect people to it," he added, "because it's flowing through our veins, our hearts, that forest water."

CORAL

Helen McGregor is an ARC Future Fellow in the School of Earth and Environmental Sciences at the University of Wollongong, Australia. Her research focuses on the El Niño–Southern Oscillation, a change in sea surface temperature and winds. Every two to seven years, El Niño disrupts rainfall patterns globally. The events initiate in the tropical Pacific and affect the way that rain falls (or doesn't) in Australia, Africa, South America, North America, and parts of Europe.

One of the challenges of studying El Niño, Helen told me, "is that we don't have particularly long records. We say it occurs every two to seven years, but we never know if it's every two or seven or five or four years, and why we get big events and why we get small events, how that relates to global warming. Are the big events that we've seen in the last decades, particularly in the eighties, were they a consequence of global warming or are they just part of the natural cycle of El Niño?"

In order to extend climate records, Helen takes X-rays of corals from the central Pacific. Corals can live for several hundred years. They have growth bands a bit like tree rings. The rings form where the skeleton is more or less dense. Some of the corals she studies are fossilized and preserved on land.

"They give us little windows into what happened during the time that they lived," Helen explained. She takes a core sample from the coral and measures the chemistry of the growth bands. Because the chemistry changes with ocean temperature and rainfall, she can use those parameters to tell when El Niño events occurred in the past.

Her research suggests that a few thousand years ago, El Niño events weren't as frequent or as strong as they are now. "The evidence is still not quite clear if El Niño just does its own thing or if actually it's responding to the broader climate system," Helen said, "but we do know that it has had phases in the past where it was quite different from what it is today." With a quieter El Niño event, Australia might have more consistent rainfall.

Helen describes climate change as "the most challenging issue facing the world and each and every one of us." At the University of Wollongong she launched Global Climate Change Week, when academics from all disciplines are encouraged to focus their teaching on climate change issues. A psychology professor might focus on how natural disasters impact peoples' social well-being. A law professor could teach about the legal protections for homeowners who build on the coast and face the challenge of rising sea levels. Helen's goal is to communicate climate science and promote a broader discussion on climate change issues outside of her expertise.

"Climate change does become very real when you're sitting out on these islands and you see just how high they are above sea level," Helen said. "They're not high at all, and it's not going to take much for them to potentially erode. We're still learning about how islands in these regions develop, so is it a case of, is the island able to keep up in a way, or redistribute sediment around and build up? Or is it a case that the rate of rise is too fast and those processes can't kick in in time? I think that's what I wonder about when I sit at the end of the day and think, all right, what are we doing tomorrow, and will this spot be here in fifty years' time?"

BLEACHING

Zooxanthellae are microscopic, photosynthetic algae that live within coral tissues. They capture sunlight and turn it into energy, providing nutrients in exchange for living inside the coral's body. When zooxanthellae are under stress—from a high ocean temperature, for example—they die or leave their host. This process is called bleaching. When the zooxanthellae leave the coral, so does the color.[11]

Coral reef scientist Lyndon DeVantier has been studying zooxanthellae for the better part of three decades. He mourns the bleaching of an ecosystem that he deeply loves.

In the 1980s, scientists documented bleaching events in the Caribbean Sea.[12] "I was beginning to do my research at that stage in the early 1980s, and those of us in the Pacific thought, *Well, that's the Caribbean. It'll never happen here*," Lyndon said.

But it did. In 1998, the Great Barrier Reef had one of its hottest summers on record.[13] Aerial surveys of 654 reefs conducted by scientists from the Great Barrier Reef Marine Park Authority showed that 74 percent of inshore and 21 percent of offshore reefs had moderate-to-high levels of bleaching. Many of these reefs recovered. Some did not.

The 1998 bleaching event prompted scientists to look at zooxanthellae more closely. It "was the first realization that actually this relationship, which wasn't even very well understood at the time, was actually quite sensitive," Lyndon said.

When the water gets too hot, zooxanthellae produce too much oxygen. The oxygen radicals poison the coral. "When that relationship breaks down to the point where there is too much oxygen for the coral to cope with, it actually ruptures its own tissues," Lyndon added. In those circumstances, the corals typically die. "They don't necessarily die outright. You may lose a large proportion of a colony, but not the entire colony. Or you may lose the entire colony," he said. "It's an incredibly complicated business."

In 1995, Lyndon led a field team that took part in one of the original surveys of the length and breadth of the Great Barrier Reef. In the years since, the reef has lost approximately half of its coral cover. While the data sets aren't as complete in the Caribbean Sea, the reefs there have also experienced degradation from climate change and synergistic impacts like disease. "Diseases spread more rapidly in warmer waters," Lyndon said. And as the oceans acidify, it becomes more difficult for corals to build their skeletons. Coral reefs are living archives. Study a coral ecosystem, and you'll find a signature of time—the telltale markers of life: We were here. We died. Reef communities have been on Earth for the best part of six hundred million years. Though some reefs have disappeared with subduction, many are still fossilized and visible—a testament to previous eras' growth. "The reason we know about it is because reefs leave a beautiful signature in the fossil record, because they build huge structures of calcium carbonate limestone," Lyndon explained.

The animals that built reefs have changed through time. Hundreds of millions of years ago, cyanobacteria called stromatolites built reefs in the world's oceans. Stromatolites still live in the waters of the Bahamas, the Yucatán Peninsula, and Western Australia's Shark Bay. They're one of the longest-living organisms on the planet.

Then came the sponges. Devonian sponges built reefs between 419.2 and 358.9 million years ago. After each of the last five mass extinctions, reef gaps in the fossil record ran for approximately ten million years. Then either the same group, a modified group, or a completely different group of animals and plants managed to recreate coral ecosystems.

Scleractinia, or polyp animals that build a hard skeleton, appeared about two hundred and twenty million years ago and have continually evolved since that time—"albeit slowly," Lyndon said. "In terms of speciation rates, reef building corals are among the slowest. Other than in some very, very unique places in the world where the physical and biological conditions have combined to provide these little areas of speciation, where we've had more rapid rates. But typically, corals are very slow."

Warming oceans and ocean acidification—the process by which carbon dioxide dissolves into the oceans and changes the pH of the water—are putting coral ecosystems under stress.

Lyndon has spent his career documenting various species of coral that live in different parts of the world and trying to use that biogeography to inform the design of a global network of marine parks. He recognizes that climate change and ocean acidification aren't going to get better in the short term. Greenhouse gases in the atmosphere will continue to heat the planet for the next few decades. "There's a commitment already locked in of continued warming, even if we were able to stop emissions tomorrow," Lyndon pointed out. "So the situation is complex."

Coral can't survive very far above the Tropic of Cancer or below the Tropic of Capricorn. Bleaching events in 2002, 2006, 2016, and 2017 have tested the capacity for acclimation within the relationship between corals and their algal symbionts. Coral scientists are trying to unravel the complexity of that relationship. Lyndon's colleagues are studying the rate at which corals can keep up with rapid climate change.

"The jury is totally out," Lyndon said. "The modeling that's been done for the next hundred years suggests that high ocean temperatures, in tandem with ocean acidification, will make most areas of the world marginal for coral survival," he added. "We may drive a large proportion of this group to extinction."

In 2008, Lyndon was a contributing author on a study for the International Union for Conservation of Nature's Red List of Threatened Species that looked at risks of extinction.[14] At that time, a third of reef building corals were at risk. Since then, the situation has declined. Coral reefs haven't been able to keep up with the rate of change of our climate system. "We are driving the system at rates beyond which these animals are typically able to cope," Lyndon said. We are conducting "an uncontrolled experiment on a large group of ancient animals."

While Lyndon loves corals from a scientific and aesthetic perspective, he also notes their utility. Reefs also provide ecosystem services to humans in the form of fisheries and coastline protection. They filter and improve water quality. And tourism is a moneymaker; the Great Barrier Reef generates over six billion Australian dollars in income through visitors each year.[15]

"They're an enormous service to humanity. But they're also a remarkable feature of the Earth's biosphere in their own right," Lyndon continued. "Who are we to be deciding their fate?"

Despite the threats against them, Lyndon remains hopeful and confident that reefs will be here into the distant future. "We should never forget that Earth is a water planet," he said, "We actually tend to forget that in our daily lives because we're terrestrial beings. But there's a lot that we can learn from the marine realm. And the corals and coral reefs can tell us a lot about where we're at and where we're potentially headed in regard to climate change."

ANTARCTIC MOSS

The flora of the Antarctic continent is limited by a climate of extremes: 250 species of lichen, 100 mosses, nearly 30 liverwort, around 700 terrestrial and aquatic algal species, a number of microscopic fungi and two kinds of flowering plants also call the continent home.[16] And they all perform feats of biology in order to survive.

When we spoke, Diana King was a PhD researcher at the University of Wollongong in Australia studying how Antarctic mosses, an indicator of climate change, are changing over time. Antarctic mosses are tiny—sometimes five hundred individual mosses can grow in a square centimeter—and they grow slowly, usually less than a millimeter per year. Antarctic mosses can desiccate readily; they're able to dehydrate and then go into a stasis mode. "Then if you give them enough water and sunlight and nutrients, they're able to come back and happily photosynthesize again after long periods of drought," Diana told me.

"They're tough, hardy little plants," she added. "As long as they've got enough water to survive, they will."

Diana and her colleagues have been able to date some of these mosses to five hundred years of age. She studies them as an indicator species. Because the ecosystem is simple and relatively untouched by humans, Diana can study the mosses as an early warning system for climate change effects globally.

From 2003 to 2013, Diana found that one species of moss, *Ceratodon purpureus*, which is found globally, has been increasing in abundance in the Windmill Islands of East Antarctica, close to Australia's Casey Station. During the decade she analyzed, the species went from being uncommon to present in about 50 percent of samples.[17] *Ceratodon purpureus* is intolerant to being submerged underwater. The increase in abundance indicated that there is less water available, likely due to changes in the winds around Antarctica.

In 2020, Diana and her colleagues recorded a heat wave in Antarctica. Temperatures soared above 0°C—accelerating ice melt. Casey Station reached its highest temperature ever documented: 9.2°C. The highest temperature on the continent was 20.75°C on February 9, 2020. Extreme events impact water availability, vegetation health, and abundance.

DANCING

In 2002, Carla Donson traveled from New Zealand to Sydney, Australia, for her thirtieth birthday. The flight was delayed because of fog. She knew, before she left, that there was a drought in New South Wales. The state hadn't had rain for months. Thunderstorms were predicted for the day of her arrival. When she finally made it to Sydney—halfway through her birthday—Carla took a bus to the middle of town.

"I watched these huge gray clouds coming over the city," she told me. "Very dark."

A few minutes later, it started to rain. "The heavens opened with the most tremendous rain I think I've experienced in my lifetime," she said, smiling at the memory.

And, because people had been so deprived of rain, they ran out of their offices. Thirty-story buildings emptied of people in business suits and dresses. "There were women kicking off their high heels, dancing in the gutters, putting their hands up to the sky," Carla remembered. "And it was incredible to just be in the middle of the city watching people behave in a way that they wouldn't ordinarily behave because they were so excited to see rain."

RISK

On a hot afternoon, I pedaled through Wooyung in New South Wales. Enticed by hand-painted signs for peaches, nectarines, lychees, and mangoes, I pulled into a fruit and vegetable stall. I propped my bicycle next to an array of coconuts. I bought a banana and a nectarine and listened as Terry, a lifelong farmer, told me the story of how floods are more unpredictable than ever before. His voice was as gruff and weathered as his skin.

"Here we used to get floods in January, February. March was your wet monsoon," Terry told me. "Now they can come any time from winter. The biggest flood we had, 2005, was the end of June: thirty-four inches in about three days."

Rain, in these parts, is security. Any farmer, planting a crop, assumes a certain amount of financial risk, hoping that the rainfall will be more or less predictable. For Terry, that risk has become too large.

"There's risk planting any crops on the flats now," he said. "You used to be able to plant watermelons, rock melons, and things on the flats and you seldom got flooded out before maybe January, February. Now, anytime, there's such a risk," he added. "Too much money involved to put many crops down on the flats."

DROUGHT

I continued to pedal north, burning under the sun, making up my route as I went. The light lengthened and the season changed. I learned to identify the smell of bat poop, sweet and pungent. In sunsets, whole clouds of bats would swoop and undulate, confident against the orange sky.

In the in-between places, I camped. In cities, I stayed with Warmshowers hosts. Heather Horne, a cyclist and senior project officer at the Department of Transport and Main Roads, was kind enough to host me in her spare room. Heather has lived in Brisbane full time since 1999. She told me that since she arrived, the number of hot days has increased. The city is hotter, drier, and "more unpleasant" than when she studied there in the 1980s.

And of course, when the land is hot and dry, access to water becomes a concern. During a drought from 2003 to 2004, it rained so little that the Wivenhoe Dam that supplies Brisbane with the majority of its water dropped to a fraction of its full capacity. Every resident was placed on severe water restrictions. "We couldn't use our garden hoses. We couldn't wash our cars," Heather said.

"I wish things were different. I wish that not only did we all take more personal responsibility in terms of our choices, what we do with our lives, how we use water, how we use energy, how we consume in general," Heather told me. "My dearest wish is that we had a federal government that took the threat of climate change seriously, because it's not some loony left-wing conspiracy."

Climate change, Heather mused, is perhaps a misnomer. It's not just one thing changing—it's everything. "It means the change of everything that we have known. We won't be leaving our children or grandchildren anything much to live with," she concluded.

In the morning, before I left, Heather made a cake and left a note with a generous slice for me in aluminum foil: "fuel for your ride."

FLOOD

Rebecca Gammon, a Canadian living in Brisbane, told me about the opposite extreme in 2011. "As a Canadian coming to Australia, I was expecting drought, because when I first moved here that's all I was ever told about," she recalled. There was a four-minute regulation on showers. Rebecca has curly hair like mine—four minutes is barely enough to get it all wet.

Then, in 2011, the water "just showed up," Rebecca told me. "We knew it was flooding up north and then all the sudden one day Brisbane was getting evacuated."

Rebecca called her mom to tell her not to worry. She put her belongings as high in her home as she could. Then she evacuated the city. Trains and busses were canceled. The king tides reached five meters. Water came up through the storm drains.

During the 1974 floods in Brisbane, people boated on some streets.[18] In 2011, people came home to cows on their patios, bull sharks in the roads, and snakes on their roofs. "All the animals were fleeing," Rebecca said.

One of her friends lived in a hard-hit area. After the waters receded, she went to help. Her friend had twenty minutes to leave and lost everything.

The water came into the house at a rate of two inches every minute. It almost reached the roof.

"The military was there and they were just gutting every house. Everything was brown," she added. Rebecca could see the line, ten feet up the trees, where the water had reached the day before.

"In true Australian fashion," Rebecca said, "there were traffic jams getting out to these areas to try to help everyone out. Every shovel was gone, every gumboot in the city was sold out, and on the street you have all these families devastated and they're still laughing, smiling," she reminisced. "People were bringing slices of watermelons to help everyone out, all the volunteers."

On the way home, they passed a boat stranded on the highway.

FLOOD

I tried to outpedal her, but serendipity followed me wherever I went. Peggy Dutton, a woman whom I met outside a grocery store in Agnes Water, took one look at my bicycle and said, "You must be traveling a long distance."

Peggy, a cyclist herself, invited me to stay with her family for the night, which turned into two. In our time together we mounded potatoes, rode our bikes to the beach, and shared stories about water birth and love. When Peggy sent me off she called ahead to her friends in the Boyne Valley—Blair and Rebecca Smith—who let me sleep in their daughter's pink room. Their nine-year-old son, Braeden, bouncing around the leather couches in his living room, told me a story about climate change: "This is a story about a very, very, very, four years later, very, very, very awesome boy!"

"How is that a story about climate change?" I asked.

"Because, silly, four years later the climate had changed!"

Braeden's dad, a former professional basketball player for the Townsville Crocodiles, told me about the exceptional number of one-hundred-year floods the Boyne Valley had experienced in the last five years.

"I think water really affects us here. We have a lot of floods and we go from drought to flood. Our winter is very dry. We have big droughts and then it breaks and we have massive floods," Blair said. "A lot of the roads get washed away." In 2013, a flood washed out the road and his family couldn't get home for eleven days.

"When we got back the whole valley had been absolutely devastated and a lot of the farmers lost their crops," he added. "Where I work in Monto, our local farmers have been flooded. I think we've had five major floods in the last three years where they've lost their crops every time. They lose their fences so it's a lot of rebuilding for people over and over again."

In generations past, intense floods might come only once every hundred years. "We've had five one-in-a-hundred-year floods in the last three years. So definitely, I guess that's the climate changing and the water affecting us big time," he said.

"But you need it, because after the floods it's going to be drought. So it's important. Interesting cycle of destruction, really. You go from plentiful to nothing to plentiful again."

PLATYPUS

I accompanied Blair's family to a music festival where I met Jack Viljoen, a musician who worked with his wife, Corna, at the Cluden Wildlife Rescue and Rehab Centre. Together they looked after injured and orphaned wildlife before releasing them into the wild. Jack invited me to pedal up the road to stay with them on my trip out of the valley.

When I arrived, it was late in the day. Jack insisted that I spend sunset by the water. "If you stay very still," he told me, "you might see a platypus."

I walked out to a dock where I lay down, belly first, scanning the creek for any signs of motion. The sky purpled. Just when the last of the orange light was thin on the horizon, I saw it: the ripples around its bill a giveaway. I stifled my urge to shout into the silhouettes of trees. As calmly as it had come, the platypus left, back to its evening activities: underwater, mercurial.

Later, I read that prolonged drought and other impacts of climate change are pushing the platypus toward extinction.[19]

Back in the house, Jack told me a story about water—how he and Corna rescued seventeen animals during a flood in 2013. The creek that I saw the platypus in rose ten meters and came through the house and the sheds. "It was massive," Jack recalled. The hills funneled the water into the valley.

"At that stage, we were pretty heavy with animals. We had quails. We had sugar gliders. We had possums. We had all sorts of wallabies and kangaroos. We had a couple of birds. We had a hell of a lot of animals, and being wildlife carers and tree huggers and all that sort of thing, we didn't worry about grabbing cars and valuables. Get that out of the way. We grabbed the animals and when the time came for us to evacuate, we went to higher ground, just south of us," Jack said.

True to form, Jack and Corna took food for the animals and very little food for themselves. They drove to a lookout and tried to pitch a tent, but a big gust of wind came through, folding the tent and breaking it. "So we ended up spending the night in the car with eighteen animals," Jack recalled. One of those was a snake—a spotted python. There wasn't any mobile reception.

The next day, emergency services arrived with a boat. They approached and asked if Jack and Corna needed help. At first Jack and Corna refused—they had enough food and weren't in danger. The first responders continued upstream and returned a few hours later, insisting that Jack and Corna come with them.

"We said, 'Well, we've got all these animals with us.' And they said, 'No, the animals can't go.' We said, 'Well, then we're not going anywhere.' And eventually they agreed that we could take the animals, and this was after a lot of negotiation. It was like talking to the United Nations about, you know, world peace or something," Jack said, laughing.

All the animals were allowed on the boat, except for the snake. "So we took the snake, made it nice and safe and everything, and got all our little animals and it looked like Noah's Ark on this rescue boat," Jack said.

"Everybody was more than happy to just grab an animal," Corna added.

"That's right. When we finally got on board all the people just grabbed a pouch with a little baby in it," Jack continued. All the animals survived except for the quails, "possibly because there was a snake with them in the back of the car and the snake got out."

A month and a half later, they were able to return home and start to clean up. In the flood they lost a colony of rock wallabies—"they moved on because of that water"—a sugar glider, a black-striped wallaby, and bettongs.

Cars and tractors were destroyed, too, but it wasn't as important. "That's just stuff," Jack said. Raising the animals from babies, they had grown fond of them.

"We keep saying that if something like that ever happens again, we'll definitely pack it up and move on," Jack concluded. One flood was enough.

VARIATION

I met Merlene Bowman in Clairview, halfway between Mackay and Rockhampton. She grows purple and red orchids in her backyard—some in pots, some suspended from a wall.

"I come from the old school. I do not believe in climate change," Merlene told me, the salty air of the Pacific around us. "What I believe is sometimes it is dry and sometimes it is wet."

Merlene told me that fifty years ago, when she was in school, "it used to rain a lot." The seasons were fickle. Sometimes the sugarcane did well, and sometimes the mangoes did. "It has just been like that all my life," she said. By Merlene's logic, the weather was variable then, and it is variable now.

I didn't press her. It was easier to talk about flowers.

BOOM & BUST

Peter McCallum has been involved with Mackay Conservation Group, a nonprofit focused on sustainability in central Queensland, since 1993. Mackay is a flat, coastal city in the tropics of Queensland. The Pioneer River runs through it.

In the past few years, Peter told me, cyclones have come close but just missed Mackay. Storm surges, high winds, and flooding in the river can all cause damage. One climate impact is clear: insurance.

"The cost of insurance has gone up by about fourfold in the past five years in Mackay," Peter said. "People in the city have noticed one of the effects of climate change is huge increases in their household insurance policies."

The job of the insurance industry is to "put a price on things like climate change. They understand the costs of the impact of major weather events," he noted. "They see that in the future there's going to be more and more intense cyclones. So they're betting that people want insurance and that they are willing to pay a lot more for it."

At only thirty-six feet above sea level, Mackay is vulnerable to sea level rise and erosion of beaches. "Lots of houses around the area will start to find that they're too close to the shoreline and that the erosion will start to take away their front yards and maybe even their houses," Peter said. "These sorts of impacts on the human population are going to be quite serious here."

Not everyone agrees. George Christensen is a member of parliament for the Dawson electorate, an area in Queensland stretching north from Mackay to the southern suburbs of Townsville. Mining, manufacturing, agriculture, and tourism are the most lucrative industries in the region. George is known for his climate change denial and pro–coal mining stance.

In 2014, he attended a conference in Las Vegas hosted by the Heartland Institute, a think tank that questions human-made climate change.[20] While there, he likened climate change to a disaster movie plot. "When we are in a flood, they tell us too much rain is a sign, more hurricanes is a sign, fewer hurricanes is a sign, the sky is blue—it's a sign, gravity—it's a sign," he told the audience. He rejects climate science and has called environmental groups "terrorists."[21]

In Mackay, coal mining in the Bowen Basin to the west has provided jobs. "There was record low unemployment because everyone was employed either in the coal industry or in associated industries. And, you know, times were great. But now the coal industry is declining," Peter told me. "The market's oversupplied. And also people are starting to move away from coal into other forms of electricity generation. So our federal member is actively campaigning to get new coal mines open for some reason, in turn to provide more coal into an oversupplied market."

ENERGY

One afternoon, I pedaled past Gladstone Power Station, Queensland's largest coal-fired power plant. I propped my bicycle outside, walked into the reception area with my cardboard sign around my neck, and on a whim asked for a story about water or climate change. I was expecting a no. But it couldn't hurt to ask.

The receptionist made a phone call.

Dave Greinke, a safety and emergency response coordinator, said yes. We bonded over bicycles. He rides his bike each day to work, about half an hour each way. He was quick to tell me that his home runs on solar power. He trained as a chemist. We sat across from each other in a conference room with stale air.

Gladstone Power Station has been operating since 1976. It is Queensland's largest power station, with a capacity to generate 1,680 megawatts—enough electricity to power Boyne Smelter, an aluminum smelter down the road. The rest goes to the state grid.[22]

Each year, around four million tons of black coal arrive at the station by rail. This coal is turned into heat energy, which heats water and produces steam. The steam turns one of six 280-megawatt turbogenerators, which then produces electricity: 16,200 volts, to be exact. Transformers convert that power to a level that can be transmitted—132,000 or 275,000 volts.

Dave was not unaware of the impact of coal on climate change. "Unfortunately, burning black coal does emit CO_2 emissions. So that does contribute to certainly the levels of CO_2," Dave told me, "and CO_2 levels have been slowly increasing, as we all know, around the stratosphere and globally, around the world."

Brown coal and black coal differ in their quality, I learned. Brown coal has more moisture and ash. In order to get the same energy output, you need to burn more brown coal "and consequently you're producing more carbon dioxide emissions."

"But obviously, energy demands always look at the cheapest way to produce it," Dave acknowledged.

Since I spoke with Dave in 2015, renewable energy has accounted for an increasing amount of Australia's electricity portfolio. In 2019, 21 percent of

Australia's total electricity generation came from renewables. In the last decade, renewable expansion has more than doubled; small-scale solar generation has grown by 44 percent and wind generation has increased 15 percent per year. Hydro fluctuates around a fairly consistent level, according to rainfall and market conditions.[23] If these trends continue, maybe—just maybe—coal will become a thing of the past.

COAL

Australia is rich in coal, both black and brown. The Bowen and Surat Basins in Queensland and the Sydney basin in New South Wales are the most significant black coal resources. Coal is Australia's largest export—each year, the country exports more than AUD$40 billion of coal to Japan, India, the European Union, Republic of Korea, Taiwan, and elsewhere. At the current rates of production, Australia could continue extracting this amount of black coal for about ninety years.[24]

Brown coal in Australia comes mostly from the Gippsland Basin in Victoria, where it is used to make electricity. Following current rates of production, there is enough brown coal to last for five hundred years. Australia ranks among the world's largest exporters of coal.

In Gladstone, I found a patch of sea and sky sandwiched between an aluminum smelter and a coal exporting facility. These operations represent jobs. These jobs have an impact. Bits of coal wash up on the beach instead of seashells. Some people swam, but my host, a cyclist I met through Warmshowers, didn't recommend it. I bent to pick up a piece of coal from the sand. Black, ridged. A fossil. The energy within: what plants absorbed from the sun, millions of years ago.

COAL DUST

Marmor isn't a coal mining town, just a village on the transport route in Queensland. The coal dust gets everywhere: windowsills, tabletops, the kitchen ceiling, even inside a home that sits a good three-quarters of a kilometer from the train tracks.

The starting salary for driving trucks at the mines out west is somewhere north of AUD$100,000.[25] Anyone would want a piece of that pie. Fly in, fly out. Live in a camp. Get the work done. Twelve-hour shifts. Reap the benefits.

I breathed in the dust while sipping tea with Ruth and Bob Hayward on their porch, and recorded a story about the cyclone that passed over their home that February, almost ripping off the roof and felling several trees. We talked about what fuels the climate system and what fuels the economy. I tracked my finger through the coal dust on the windowsill.

One of my hosts used to manage service stations out west. The miners would come in after work to buy a cola, still in their work uniforms. Save for a ring around their eyes, they would be covered in the black stuff.

One day, one of the workers changed shifts and came in with his face clean to buy a cola. She didn't recognize him.

CYCLONE

In February 2015, a cyclone flooded Ruth and Bob's home in Marmor, Queensland.

"For a normal thunderstorm here, in the wet season, it's not unusual for it to go for half an hour and get two inches of water," Bob told me. "A lot of water floats down that street. Everywhere is just covered with it. And then three, four hours later, when the sun comes back up again, it's all over. You'd swear nothing had even happened."

The Haywards weren't expecting a cyclone. And they hoped to not see one again. "Believe you me, I don't want to go through another one," Bob said. "But that's climate change."

"We're not tree huggers," the Haywards were quick to add.

Ruth noted that the cyclones are moving down the coast more frequently. "Southeast Queensland's been copping a lot of stuff that it didn't used to get," she said.

"And they're fierce. More fierce now than ever," Bob added. He pointed to his neighbor's roof. "In a cyclone where water's probably coming down at about three feet of water an hour, hitting your roof with cyclonic winds pushing it, can you imagine the stress that's on that roof?"

One neighbor, Ginny, lost half of her roof. Some of it landed on the Haywards' roof. Mari lost her roof in its entirety. One house down the road was wiped out completely.

"My shed door ended up from the paddock right there into the garden there. Airborne," Bob said, chuckling.

In their backyard, Bob and Ruth keep a menagerie of birds. Ruth notices the dry days because she has to fill the bird baths two or three times.

After the cyclone, Bob was surprised at how quickly some trees regenerated. "Within a week, it's sprouting again," he said.

"But we need more rain," Ruth said.

"No, we need it spread out," Bob laughed.

A coal train passed in the background. I started to count the carriages and lost track after ninety. Ruth swirled her spoon around her teacup. Over it all, her birds sang.

RUNOFF

Tara Hayward, Ruth and Bob's granddaughter, works as a nurse. Her boyfriend travels to work opening coal mines. We spoke in her kitchen, where multiple fish tanks bubbled, casting a blue light on her face. She had a yellow pet snake curled around her arm.

Tara wrestled with the impacts of coal mining on places like Dysart, a mining town northwest of Rockhampton, where runoff from mining has contaminated drinking water. She has seen the health impacts.

In 2013, "all the water went pretty crappy," in Dysart, she told me. At that time, she worked at the hospital.

"For about three weeks over the Christmas period, we had no water whatsoever," Tara recalled. Potable water arrived from nearby cities in trucks. The water from the tap "was about the same color as a dam. Real muddy. Murky. You couldn't drink it. You couldn't bathe in it. Even the hospital, we had nothing."

The hospital had portable toilets installed. "We had a massive amount of diarrhea, really. Which is very unfortunate because we had no water," Tara added.

Washout from the mine had decimated Dysart's water supply. "It just destroyed the water. And the water got that low because there was no rain. It just kind of brought up all the crap, you know, all the coal from the bottom of the dam. Still not fixed. It's still really bad. You can't drink the water out there. It just tastes terrible and it's very unhealthy."

DREAMTIME

Aurore is an Aboriginal Australian who belongs to the Dja Dja Wurrung people. Her traditional lands include the Loddon River in central Victoria. "The story I want to share with you is a very ancient dreamtime story," she told me.

Kuparu is the word for koala in the local language, Woiwurrung. Kulin are an alliance of five Aboriginal Australian nations.

"When time was only young," Aurore told me, "we would hunt all sorts of food, and we didn't mind hunting Kuparu. Now Kuparu didn't mind being hunted, but Kuparu did object to the way we cooked him."

One day, in anger, Kuparu took all the water. The people had no water left. So the people went to speak with Bunjil, the creator of all things. The creator approached Kuparu and asked why. Kuparu said, "because Kulin does not eat us the right way, does not share respect with us."

And Bunjil said, "How should Kulin cook you?"

And Kuparu said, "If Kulin promises not to skin us before he cooks us, then we will return all his water to him."

Kulin decided that was a good enough thing to do. They promised from that day forth never to skin Kuparu before cooking—to always cook a koala with the skin on. And then they got all their water back.

ZIP TIES AND MAGPIES

Thoreau once wrote that river towns have wings. In Rockhampton, I learned what it meant to be swooped: a beak suddenly inches from my ear, the overwhelming sound of feathers moving air. It was a warning. I was intruding: this bird did not want me around her eggs or newly hatched chicks.

I brought it up with a Warmshowers host, who suggested affixing zip ties to my helmet, pointy end out. It worked. The magpies left me alone. Even after swooping season, I kept the zip ties on. It helped deter the wrong type of people, too.

SUGARCANE

When I say that the sunsets in Sarina smell like sugar, it's not a metaphor. The mill runs constantly in this small sugarcane town in central Queensland, spewing out char and molasses and rum.

I arrived in town before dusk and wasn't keen on camping in one of the caravan parks—they tend to be hostile places for a solo female. I much preferred to sleep alone in the roadside wild. I followed my nose up a hill, searching for a good, flat spot to stealth camp. I paused in the middle of a residential road to think and to write. Then I felt two sets of eyes to my left: two kids playing in their yard had paused to take me in—the bicycle, bags, ridiculous magpie-deterring helmet, neon vest, the neck of a guitar sticking out the back.

I waved, and the kids waved back. Across the street, children ran circles on top of a trampoline, bouncing in the golden light. A neighborhood cat crisscrossed the road. There were no cars in sight.

"Hi!" I said to the kids. "This is an odd question, but do you know of somewhere where I might be able to fill up on water?" I was empty. Camping without water is difficult.

The girl with pigtails looked up at me. "We have a hose you can use!"

We talked as the stream of water jetted into my plastic jug. School holidays were coming up, she told me. This is a small town where everyone knows everyone else. A long time ago her family used to live in Melbourne, but she doesn't remember that. Her brother watched silently. Then their dad called them in for dinner.

"I have to go," the girl said. "Good luck." Across the street the trampoline was empty. I rode on, searching for trees and flat ground, the gift of the water heavy behind me.

Eventually I found a playground surrounded by houses that might work as a campsite, but what if one of the neighbors objected? A woman was watering her plants nearby.

"Can I ask you a question?" I asked, cycling closer.

"Of course!" she said.

"Do you know of a place where I could camp around here? Would your neighbors be okay with a cyclist sleeping in the playground?"

She thought for a second, hose in hand. The water pooled in the grass.

"The neighbors are great, but you should really spend the night with us," she replied. "Come set up your tent. Have you eaten?"

And that's how I met Tina and her husband, Gus, and her two kids, Chloe and Casey. Chloe was going into tenth grade and Casey was in elementary school. They had guinea pigs and a rambunctious puppy and a cat named Doris.

Gus suggested that I spend an extra day in Sarina to learn more about the coal industry, and it was an offer I couldn't refuse. I had been cycling next to railway lines where up to forty trains passed per day, 125 carriages each, full to the brim. Most of it goes from the mines straight to the sea, where it floats to markets in Asia and beyond. The scale is mind-boggling.

Tina and I drove to the coal terminal at Hay Point, one of the largest coal export ports in the world.[26] You wouldn't know it's there unless you were looking for it—the terminal is well hidden by the surrounding hills.

It was mesmerizing to watch the flurry of activity from above: big trucks, looping trains, cascades of water sprayed from the nozzle of fire-engine-size hoses to keep the coal dust down. Tina pointed out blips on the horizon—each a ship lined up at sea, waiting.

I fixated on the spray of water bending light into rainbows near the hive of activity. Even from a distance, I could hear the trucks reversing and revving their engines. Something large was being built.

I found out exactly what it was from a storyteller in a sundress. Her husband was on contract to build two dams for the ever enlarging coal terminal. "What could they possibly need all that water for?" I asked.

"To spray the coal. To keep down the dust." The woman went on for a bit about how the "bloody greenies" had been protesting these dams for years, but they finally got the projects. And now her husband had good work. It's a moneymaker, that black stuff.

CROCODILES

I met Rhianna Steindl, a wildlife ranger, on a rainy morning in Bundaberg. Rhianna licenses wildlife and manages crocodiles throughout the state of Queensland. Sometimes, this means getting in a boat and putting a spotlight in the water, in hopes of harpooning a croc.

Whether climate change is impacting the movement of crocodiles in Queensland is an open debate. There have been historical sightings of crocodiles in the Mary River since the late 1800s, but never in large numbers. Climate change creates an opportunity for crocodiles that are displaced by population pressure in the rivers of northern Queensland to migrate and repopulate southern river systems. Warmer water temperatures support breeding and egg hatching. Confirming their movement would require putting GPS tracking devices on juvenile crocodiles in rivers like the Fitzroy at Rockhampton to the north to see if any of these animals expand into southern waters over their lifetimes.

The rangers confirmed their first croc sighting in May 2013 and named him Marc, short for Mary River Croc. It took two years to trap Marc; he was devious. They tried traps and harpoons. "We spent a fair bit of time on the water in our boat. We started off with a little boat called *Cormorant*, which is probably a foot or two off the water," Rhianna said. "You sit out in the middle of the river going, 'Well, I sure hope it doesn't really like boats.'"

Then, in July 2013, a second croc turned up.[27] The rangers called him Scutey, because when they scrutinized photos, they saw that the crocodile had been scute-marked by researchers farther north. Scute marking is an identification system that involves cutting plate-like scales on the croc's tail with a pair of scissors, a knife, or a scalpel.

"Unfortunately, even after researching various projects, we couldn't figure out which crocodile farm or research project this animal may have originally come from or how it ended up in the Mary River," Rhianna said. They captured Scutey that November. He was 3.8 meters long, the same size as the boat.

Over the time her team had been observing them, Rhianna noted that the crocodiles in the Mary River behave differently than ones up north.

"The ones up north are quite territorial. They'll approach the boat. They'll have a go at you. Mainly because there's so many, and they're competing for space." In the Mary River, the crocodiles "got to know the sound of our boat and they'd be in the water a good six, seven hundred meters away," she said. Initially she could guarantee a sighting on every outing. But over time, the crocs learned the sound of the boat and would disappear underwater for a long time, outwaiting their would-be captors.

When the team caught Marc, the four-meter-long titan, "It was a pretty sad day for us because we know that we can't release him back to the wild. They go to croc farms where it's up to them, how they use them."

Both Scutey and Marc went to Koorana Crocodile Farm in Rockhampton where they were used for breeding.

TSUNAMI

David and Sharon Cox, a couple from South Africa, hosted me through Warmshowers—I was their first guest. I detoured inland from the Gold Coast to Willow Vale, Queensland, to stay with them for a few days. David is a water engineer. Sharon sells options on the US stock market. But they wanted to tell me a water story about Boxing Day 2004, a holiday they spent in Sri Lanka with their young boys, then ages six and eight.

Sharon, before leaving, had an uneasy feeling about the trip. She had a recurring dream where she was standing, looking toward the sea, holding the boys' hands, when an enormous wave came. She disregarded the dream, but a couple of days later it happened again. She woke up thinking, what would I do if this actually happened? She reasoned that she would grab the boys and run.

On December 26, 2004, the family went for breakfast at the side of the pool at the beachside hotel where they were staying in Sri Lanka. The boys ran around the pool, playing with water guns they had gotten for Christmas.

And then something like high tide came—a bit of water swept over the lawn and put mud in the pool. The older boy turned to his dad and asked him to take a photo so that he could show his teacher at school how the mud went into the pool. David obliged. He went on the beach to take the photo.

"And I was looking around and then suddenly the water really started coming over quite, quite a lot, you know, around my ankles and everything and started really building." Then the water came up to his knees. "I was looking for Sharon and the boys and I couldn't see them."

David moved into the restaurant area. The water started lifting up the tables, which were made of a heavy hardwood. He jumped over a low wall in the restaurant and went around back, realizing that this was stranger than a high tide. It was a tsunami.

"It wasn't like a surfing wave where it's a big curly, curved wave. This was more like a wedge that just grew and grew and grew," David recalled. He passed a side room where a woman told him to come inside, but it didn't feel right. He felt like he was getting closed in. So he walked out. The water continued rising, about ten centimeters every thirty seconds. Then water burst through two windows and blew out the glass next to where he had just been standing.

"There were trees and bottles, thousands of bottles and fridges and things coming around me. And then the water was rising. I couldn't stand anymore. So I started swimming up the side of the building as the water rose," David said. Inside the building, a woman in a sari screamed. She couldn't come out and David couldn't get in.

A white car lifted up from its parking space and floated off. Across the street, a blue car lifted up and then was parked in the branches of a tree.

"The water was so muddy and brown," David said. "You couldn't imagine it." The ocean was like a river, a raging torrent smashing everything in sight. He started looking for Sharon and the boys.

"I know Sharon is a good swimmer," he said. But he thought at least one member of the family would be lost, if not more. "She can't save two kids. You know, you can only save one. And so I was just watching this and just hoping and hoping it was going to stop because it just seemed to go on and on and on," he continued. "There was no way we could all have survived. It was too massive."

After something like twenty minutes, when the water was almost at the roof of the building, it started to recede. The blue car dropped out of the tree. The woman in the sari who was trapped in the building was able to wade out. A man dropped out of a coconut tree and told David to come with him to his village, inland, to get away from the sea.

On the way, they passed the chalet where they had been staying. There was nothing left of it—the structure was there, but the windows, doors, and everything inside was gone. "There wasn't even broken glass because it had been swept so clean," he said.

David started looking for bodies, fearing the worst. The power was down. The railway tracks had been lifted up. Every person he passed, he asked, "Have you seen a woman, two children?" Occasionally somebody would say yes, but it was always the wrong family.

When they got to the village, everyone was crying. The man's dad had been killed. The hotel next door had lost almost their entire staff. David apologized, but he had to leave to find his family.

He ran back toward the hotel and heard a motorbike coming through the bush. David flagged down the driver and jumped on the back. "I said, 'Where are we going, my friend?' Because I didn't care where we were going as long as I knew we were quicker on a motorbike."

As they were driving toward the police station, they saw a group of people whom David recognized from the hotel. He jumped off the motorcycle and started walking back toward the sea.

"And I kept asking these people from the hotel, have you seen Sharon with the two little boys? And nobody had seen them. I thought, *Oh, my god, you know, this is it*," he said. "*They haven't made it*."

Then one woman brought good news. She recognized Sharon's sarong pants—a baggy orange pair. The woman said she had seen her. David asked how many: one or two or three? The woman said it was her and the two boys. David was washed with relief.

When Sharon looked at the sea and saw the horizon line lifting, she knew she had to run. She grabbed one boy in each hand and sprinted. They crossed the coastal road and ran alongside the river. A man came along and picked up the younger boy and they ran along the railway line, past thorns and broken glass. Not once did the boys complain. They arrived at a village and found a roof. Then they heard about a Buddhist monastery farther up the hill. They decided to move again. A lot of people started to congregate there.

"A woman came to me and she said, 'Are you Sharon?' And I said, 'Yes.' And she said, 'Don't worry. I've seen your husband. He's fine. He's still got his glasses on.' " Sharon laughed.

"I knew he hadn't been swimming that much," she said.

They found each other, and the four of them headed for high ground. After that, the only thing left to do was to get home.

5

AT SEA

FLYING

In March 2015, an environmental activist in New Zealand introduced me to Chris Watson, author of *Beyond Flying*, a book of essays that interrogates the necessity of air travel.[1] Chris and I met up in Wellington for a flat white.

"How many flights have you taken since starting this journey?" he asked.

I tallied six on my fingers.

"The environmental footprint of all that flying will outweigh any other actions you can take," he said.

Two people on a tandem bicycle passed us on the street. I tucked my hair behind my ear and stared into my empty cup. I didn't know what to say.

That night, I calculated the tons of carbon dioxide emitted by the sum of those flights using a calculator online.[2] Chris was right. The long-haul flights far outweighed riding my bicycle or choosing not to eat meat. Any decision I made could not override it: flying is by far the worst thing that many of us do for the planet.

If I'm traveling to collect stories about water and climate change, I might as well do my best not to add to the atmospheric problem while I'm at it, I reasoned. Not one to shy away from a creative challenge, I decided to see how long I could last without flying. A year seemed like a good, if inadequate, start.

The only problem was that I was in the middle of a whole lot of water. You can divide the Earth into the land hemisphere and the water hemisphere, and New Zealand is right in the middle of the water hemisphere, surrounded by ocean on all sides.[3] I did some research online about how to cross the Tasman Sea by boat—it's one of the wildest stretches of ocean in the world, weather-wise, so sailing was out. I found that a German container shipping company, Hamburg Süd, takes passengers. It's expensive, though—about US$200/day.

It wasn't until I cycled all the way up Arthur's Pass, one of the steepest hills in New Zealand, that I had an idea. Along the way, cars rolled down their windows to cheer for me. "You can do this!" they said.

When I stopped at a gas station to stock up on snacks (gummies shaped like vampire teeth was the only thing I could stomach), the cashier told me: "If you can't make it, just roll back down the hill and we'll drive you to the top."

But I was stubborn. I ate my gummy vampire teeth. I made it. And I didn't forget the cashier's words of support. I picked a flower at the top of the mountain pass, put it behind my ear, and did a happy dance. If I could be brave enough to get all the way up and over a continental divide, pushing a bicycle full of gear, then I could most certainly be brave enough to ask for help to get me aboard that cargo ship.

That night I went on Kickstarter and set up a funding page, offering to write poems and letters and postcards from the ship to people who helped me out. I was nervous, but my support network (including folks I had never met who believed in my project) came out in full force. In a day and an hour, I exceeded my goal.

As with most good journeys, there are no straight lines. I ended up crossing the Tasman Sea twice by cargo ship, crowdfunding both legs, and almost hitched a ride to Indonesia on a yacht. In between these ocean crossings, I biked twenty-one hundred miles up the east coast of Australia, from Melbourne to Townsville, and added ninety stories to my archive in progress.

When considering cargo ship travel, it's best to schedule a passage as far in advance as is possible. Flexibility is a necessity. The original ship that I was going to take, the *ANL Bindaree*, was moored unexpectedly off the coast of California due to a mechanical problem. I reworked my travel dates by a few days to take the *Spirit of Shanghai* instead. Cargo ships can be unpredictable beasts.

CARGO

At any given time there are thousands of cargo ships in the water, carrying all manner of things nestled within shipping containers. MarineTraffic.com lets you track the movements of cargo ships as they swarm, caterpillar-like, over the planet's seas. An estimated 90 percent of all goods are transported this way. Sneakers from China, potash from Belarus, T-shirts from Bangladesh, and bananas from Ecuador float between continents, propelled by a diesel engine that's larger than a house.

In 1956, the shipping container changed the face of how we conduct global trade, introducing Lego-like, standardized units in place of a more haphazard arrangement.[4] In earlier times, smaller loads were jumbled together into a cargo hold; the loading and unloading process was lengthy and cumbersome. Now, the containers—forty feet long, eight feet wide and roughly eight feet tall—are colorful and stack together.[5] Nestled in my cabin overlooking the deck, I counted three shades of gray, plus blue and orange and red and green and white and maroon. Several gantry cranes—yellow with striped legs—rolled down the length of the hull. Draped cables slid in and out of view: an overhead ballet. The alignment of each container must be perfect. At port, the stack on board grew higher: piece by colorful piece.

I peered out my rectangular porthole, mesmerized, as above me a container crane operator peered through a glass floor, using two joysticks to carefully align, pick up, and release the containers, one by one. Each movement was punctuated by a loud, resounding *thunk* as the containers connected together. With each release into the cargo hold, I could feel the whole ship shake. This routine continued for hours: mechanical and human and messy and organized, all at once.

I found relaxation in the steady monotony, the order of it. Each container has a unique identifying number for tracking. The shape and size and corner fittings of the containers are standardized, worldwide by the International Organization for Standardization.

Later in the day, with everything arranged, we departed. I stored my belongings in a small cabin with an orange couch, a single bed with white sheets, and a laminate coffee table. The cabinets locked and all of the furniture was bolted to the floor.

While we motored out of port, the third officer, a crew member from Myanmar, led a safety tour. In case of emergency, a panic alarm—seven short blasts followed by one long one—sound. Everyone would meet at the muster station and then go to their assigned lifeboat. There were lifeboats on both the port and starboard sides that could hold thirty people each, and another two for six people. Inside, we could find fifty life vests and enough rations to survive for a while—hopefully long enough for someone to notice the lifeboat on the high seas. We practiced zipping on neoprene orange survival suits, meant to protect us from the cold water should we be submerged. In a fire, everyone would be summoned to fight the flames, save for the passengers, who would report to the navigation bridge. This, more than anything else, reminded me that I was cargo.

"Have you ever had a fire during your time on the ship?" I asked the man leading the safety tour.

"No. Luckily I didn't have a fire, but every week we have a fire drill," he said, pointing out the CO_2 fire extinguishers on each deck. He added that the engine room had a foam extinguisher. Many doors on the ship were magnetic in order to isolate the flames. There weren't any doctors on board, but the second officer took care of medical equipment. And the crew could buy beer from a store on board.

"Is it good beer?" I asked, but his walkie-talkie drowned out the question—communication from elsewhere on the ship.

All the fruits and vegetables for the galley were stored in a refrigerator nearby, which he opened for us to peer into for a few seconds. The cool air met my nostrils. A few decks away, the third officer opened the door for the crew recreation room, where two guitars balanced on the sofa. Every cabin was outfitted with a life jacket, too.

We ended our tour up on the navigation bridge. "When we are departing from the port and arriving at the port, it's very busy," the third officer said. It's best not to bother anyone who is working, he added. "But when we are going far from the port, you can come." Over walkie-talkies, the captain communicated with the port staff, navigating a safe passage into the open seas.

CREW

The first thing to know about a ship is that it is not one thing. It is many people. A crew. Stacks of anonymized, standardized containers. The port staff. The machinery. All of them: engines of a global economy. At night, the refrigerated containers glowed. I imagined what they might carry—oranges or ice cream or meat—though I did not know. I asked, but no one on board knew either. The containers were just that—units, blocks to be moved from place to place.

Even a massive ship like the *Spirit of Shanghai*—nearly three football fields long—is still subject to the pitch and roll of the sea. Everything vibrated. The ocean felt vaster than I could have imagined: horizon every which way, so many subtle gradations of blue. The weather was calm for two days, but on the third day we caught the outside of a low-pressure system. The weather churned up the waves. At breakfast, the captain instructed me to keep all of my belongings locked up in the cupboards of my cabin in case the sea went wild.

The captain, Marius Banica, was Romanian and gruff. One evening, over split pea soup, he told a story about the two pirate fishing boats six years before that chased his container ship in the Indian Ocean, about a hundred miles away from Mumbai.

The crew took photographs using binoculars. "They had guns. Their intention was clear: coming on board and capturing the vessel," the captain said.

Fortunately, the water was flat and his ship was able to outrun the armed pirates—only nearly, though. The container ship was cruising at 22 knots and the pirates at 21 knots. The chase lasted for two hours before the pirates gave up and turned around. If they had taken the ship, they would have locked the crew in a room and stolen the goods on board. At that time, the captain told me, they did not have an armed guard on board.

"We were lucky that time. We escaped," he said.

Ships are funky social spaces. There's the obvious: you can't leave. Crew members work eight-month contracts, seven days a week; officers sign on for four months at a time. The clock leaves little room for spontaneity.

We had assigned seats at mealtimes, which was perhaps my least favorite part of the whole ordeal. There were two symmetrical dining rooms—one for officers (on starboard side) and one for the crew (on port). In between was the galley (ship terminology for the kitchen). The food differed by dining room, too—the crew was served spicy noodles and rice. We had potatoes and meat without seasoning. I sat at the captain's table prodding at a baked potato while the one other passenger, a thirty-three-year-old man from Melbourne, told me he didn't believe that women should work outside the home. As you can imagine, we got along like a ship with an engine room fire. On the third night of the six-day journey, he got seasick: divine intervention.

Alex, the Russian chief officer, was more pleasant to chat with. One night after dinner he brought me a bowl of chocolate ice cream from his personal stash. While we ate, he settled into a story. Years before, Alex worked on a timber carrier. The timber was carried in packages in the holds, with other cargo affixed to the deck. Their ship carried timber from Russia to the Mediterranean Sea.

"One day we received a message from the company about another vessel in our company. I was on the bridge, on watch. The chief mate came on the bridge and read this message," he continued. "I don't know how it is in English. In Russian it's 'wave killer.' It means, at sea, when there is swell from one direction and maybe another swell in almost the same direction, the waves meet and become stronger."

The chief mate read the message, but he didn't take it seriously. "He said, 'Ha ha, wave killer,'" Alex recalled. Moments later, a high wave came and went through the deck, removing all the cargo. *"Bam,"* Alex said.

There was nothing they could do except watch their cargo sink and bob away.

Alex, a man in his forties, repositioned himself on his chair and took another bite of ice cream, which was starting to melt and pool in his bowl. Earlier that day on the navigation bridge, he told me that a release mechanism would unlock the towers of containers if the ship rocked side to side at more than a certain angle, in order to prevent the ship from capsizing. Where there is one big wave, there is likely to be another.

"The other one I saw was also on a calm day. Calm sea," Alex continued. "Nice weather. Sun shining. The swell was maybe three, four meters—for our vessel it was okay. I wrote something in the logbook because it was almost noontime. I prepared everything and signed for the change of watch."

Then, the galley cook called him. "Alex, what are you doing?" the cook asked. "What happened outside?"

"I said, 'I don't know. What happened?'" Alex told me. "The guys told me when they came downstairs in the mess room, it was full of water."

The weather was nice and the crew had opened all the windows to feel the breeze and relax. Then, a wave came in through the window.

Alex laughed and put his hands above his head. "So this was two kinds of wave killer."

I eyed the round portholes in the mess hall, imagining the wave, a sloppy surprise.

"Did you have to clean it up?" I asked.

"Oh yes," he said.

The ship listed slightly to starboard side. His ice cream spoon rattled in its bowl.

SPIRIT

The *Spirit of Shanghai* is 833 feet long and can hold 3,630 containers, each the size of a tractor-trailer's load. My bicycle, dust-mote-like in comparison, was tied up with rope in a room used by extra crew during crossings of the Suez Canal.

"Never call the ship a 'boat,'" the captain chided me. "Bad form. She is a ship."

I was the only woman on board the *Spirit*. The twenty-one crew members were mostly from Myanmar, though the captain was Romanian and the chief engineer was Russian. The chief engineer loved conspiracy theories and wouldn't stop talking about the might of the USSR and the beauty that is huge American cars.

Part of me wished that there was another female or an American on board so that we could have that false pretense of commonality (gender or nation) as a conversational starting point. But on the ship, I had hours to learn the contours of my own mind. Social hours were confined to meals. There was no internet. Reading made me queasy.

I spent my afternoons on the navigation bridge, the highest point in the stern. In between monitoring various dials and instruments and beeps, Alex told me that the ship was made of flexible steel; if it were not flexible, it would break. One job on the deck involved periodic welding to preserve the ship's integrity. In rough seas I could see the hull move across multiple dimensions—up and down and side to side, waves of motion propagated throughout the metal.

The sky over the ship and its containers felt porous, ancient: it had been places—like worn jeans are a map to the legs that wore them. The night skies were especially sumptuous. One morning I woke up just before dawn and looked out my porthole to see stars on the horizon, bright and free—light kissing the wingtips of the sea. Ten minutes later a faded, more defined blue was sprinkled with the vanilla pinpricks of celestial bodies. The sea is a place of gradual changes, of sputtering details that knocked me over and pushed me back up again.

Even with the ship's uncompromising routine, the onward surge of hours and their assigned tasks, I had the feeling that anything could happen. Water could press through our portholes. A violent storm could take the shipping containers as ransom. Waves are unforgiving. We were at the mercy of our aqueous environment. When we passed into a new time zone, the PA system announced the change in hours. But somehow, the orderliness of the ship felt superfluous. It gave shape to our days, yes, and the rotation of the watch kept us all alive. But if the sea wanted to take us, she could have. The lingering threat was always there.

Every day at eleven a.m., Alex filled up a pool in the gymnasium for exercise. It was a small, sloshy affair—sea water, the lot of it—that he emptied before lunch at noon. I almost jumped in but decided it would be even more improbable and ridiculous to try running on the treadmill that was affixed to the floor next door. I made it for five minutes, swaying slightly with the lateral motion of the boat, before one wave almost knocked me sideways and I gave up. I settled for a one-woman dance party in my cabin. (Moral of the story: running on water is quite difficult. Dancing is only slightly easier).

Over the days of the passage, I came to think of the ship as a dinosaur. It certainly roared and rumbled like I imagined a prehistoric creature would. And, like the ocean and the waves, the cargo ship is temperamental. It has moods. On board, I rocked and rolled and tumbled. I can report with great certainty that I have no sea legs to speak of, and that my propensity to bump into stationary objects while aboard the *Spirit of Shanghai* was, in fact, quite funny. Stairwell railings never felt so essential. I clung to them as I ascended and descended the floors, never knowing when a large wave would come and flatten me.

One of the crew, a big, burly guy from Myanmar who worked in the engine room, called me "strong" at dinner one night. This made me smile.

WORK

One afternoon, as the sun slanted into gold, I met a junior electrician from Ethiopia outside the crew dining room. He couldn't have been more than a few years older than me. Seeing my "tell me a story about water" sign, he invited me into the dining room to chat. We sat next to a big bowl of leftover fried rice. The ship listed gently to starboard, then to port. Beneath us, the engine hummed.

"What made you want to come to sea?" I asked.

"At some point you have to face life," he said. "You have to work."

He gazed thoughtfully at his hands.

"Working at sea has its advantages and disadvantages," he continued. "Advantages? It's very simple to get paid good money. Now I'm junior. I'm not paid that much. But after becoming a senior electrician, a much bigger salary is waiting for me. But it also has a big disadvantage. You are far away from the family, maybe your wife, your brother, your sister: everyone you love. You have to deal with it."

Standard contracts on cargo ships are nine months on, three months off. The ship travels from New Zealand to Australia to the Panama Canal to the United States, back through Panama, and to New Zealand again. One trip takes around seventy-two days.

At the end of the contract, the cargo shipping company flies the crew member home from whichever country they happen to be calling at. Each shift is between ten and twelve hours long, seven days a week. Constant attention is a necessity, and in the off-hours, there's not much to do. Crew members watch DVDs of movies in their cabin or play guitar in the lounge or sleep.

"You have to deal with it," the junior electrician repeated. "To get something you have to lose something. I like where I am right now. I made a decision to be here."

CAPTAIN

In the forty years since he became a seafarer, Leszek Daworski, captain of the *Spirit of Singapore*, had seen changes. He estimated that 30 to 35 percent of that time was spent at home with his family in Poland, but for the rest of those years, he was on the water.

"During this time I've observed a lot of changes in climate, oceans," he said. "I can't say that it's different," the captain added, but "I can say that it's not what was before."

"What kind of changes have you noticed?" I asked.

"Not only at sea but also at home, you see, the climate has changed. The seasons. When I was young, for sure, wintertime was very clear," he said. There was snow on Christmas. "Now, white Christmas is in that song only. Not in reality."

The timing of the summer season has also shifted. "Now, when we have that beautiful weather, it's in May or June." July to September, he told me, could be unseasonably cold.

"I don't have statistics, but I can observe what it was like when I was a young child and now," the captain added. "At sea? The current, hurricanes, and cyclones are different. We have a chart for typical weather according to the day, records from the last twenty years. Now, even the Miami Hurricane Center gives forecasts for six hours," though sometimes they are inaccurate.

"Nobody knows. The currents also are changing," he said. "It has a big influence on our life."

Leszek said that for a few decades, nobody noticed the changes. "When my parents were younger, things were different. And my parents are eighty-six years old. They can say something," he said.

"Most old people say now the world is terrible. When I was young, it was beautiful. No. The world is at all times beautiful, but only if you understand and accept this."

LAND

And then there was the trip's end—approaching Sydney Harbor. If I looked over my shoulder I could see the glimmering lights of the Sydney skyline, a welcome change of scenery after so many days of waves. The city looked massive, but the Tasman Sea was larger.

On board, a customs official stamped my passport. I rolled my bicycle down a set of steps, waving goodbye to the crew member who carried my panniers. A van deposited me at the road. After so much vertical and lateral movement, I didn't know what to do with my legs.

PASSAGE

Keeping the promise not to fly was not an easy thing. As I cycled farther north up the east coast of Australia, I had to get creative. My visa was running out. No cargo ships took passengers along that route. If I wanted to make it to Southeast Asia, I reasoned, I would have to hitch a ride on a sailboat.

I started posting handwritten signs at ports as I cycled north, trying to will a sailboat into existence. I left my email address and snippets of my story in ports.

"I am excited to see the water, willing to work any shifts, am a knowledgeable cook, and I am happy to clean," I wrote. "I want to learn navigation, sail handling, overnight watches, and how to handle the boat on my own."

No one wrote back.

BEACH

Looking backward, it's easy to trace the path of spontaneity—how one trip to the grocery store led to a conversation that led to a friendship, a birthday party, a sailboat. But at the time, I was just following what felt right.

In the town of Agnes Water, Queensland, a woman named Peggy Dutton approached me in the parking lot of FoodWorks, a grocery store. I was coming out, food in hand, to restock my bicycle. Peggy, on her way in, saw me with the bike and said: "You must be going a long way!"

We started to talk. Within five minutes, she had promised to introduce me to her friend, Bless Beyers, who lived a few days' ride north in Airlie Beach. (True story: I was afraid that after university I would never meet friends. Clearly I was mistaken. Great people are everywhere, especially by the sea. The hardest part is often saying goodbye.)

For my twenty-third birthday, Bless threw a party on the beach. She and her friends filled my birthday with water and music. We made a nest-like structure to sleep in out of branches. Someone brought a guitar. We played music and swam out to a shipwreck and watched the sun set, falling asleep in the cocoon of each others' joy. We must have summoned something that set the magic in motion—the delicious improbability of it.

The morning after, I woke up at first light. No one else in the nest was awake. Not wanting to disturb the sleepy revelers, I decided to gather water for oatmeal and coffee. One of Bless's friends had a dog named Littlefoot. He followed me out of the nest on my quest to find the beach's public water fountain, where I filled up two empty jugs.

When I returned to the nest, water in hand, someone walked by. I traced her footprints backward in the sand, out toward the marina. Littlefoot approached her and put his paws on her leg. I apologized for his friendliness.

"You have an American accent!" she said, scratching behind Littlefoot's ears.

"So do you!"

We walked and talked, nestled into the comfort of some kind of commonality, here at one edge of the earth. I learned that her name was Jenny Ramirez. Jenny told me that she and her partner, Randy, were helping Bruce Albert, a retiree, sail up the east coast of Australia to the Philippines. They were stopping on the way in Indonesia.

I told her that I was looking to catch a ride to Indonesia. We made plans to meet up in a week at the Townsville marina, where they would be stocking up to resupply. I couldn't believe my luck: beautiful and wild and tumultuous like the sea. It all started in a nest. (Dreams always start in a nest.) When it was time to leave, I said goodbye to Bless and pedaled north.

SAIL

Jenny and I met on the dock at Townsville. She walked me down the dock to see the monohull we would call home: *Far Fetched*—a suitable name for how preposterous it was that they agreed to take me and my bicycle on board in the first place. I left behind Townsville, the coast: its bats and rivers. I was porous: eager to learn.

Sailing has its own language and movement. Jenny was a patient teacher. The toilet is the "head." The kitchen is the "galley." One minute = one sixtieth of a degree of latitude = one nautical mile. Every rope has a name, and none of those names are "rope." A winch is different from a wench.

Before dusk the lot of us walked the length of the surrounding docks to scope out the other vessels. I learned about the difference between a catamaran (two hulls, less tippy initially but totally out of luck if it tips over) and a monohull (which could, in theory, right itself if flipped). Being inside the hollow of the yacht—forty feet long and thirteen feet wide—felt like floating inside a tree. The hatches let in light from the sky. I liked that feeling of porousness. I loved looking out of the rectangular windows and seeing water. The water we drank came from plastic jugs, to be carefully conserved.

I offered myself to the ocean's arms and trusted this crew completely. I had to. Once we left land, the only laws were our own.

The beginning of a sailing trip involves a lot of waiting. At night, I watched the reflection of the marina lights on the water. So many masts. We waited for the arrival of a new gasket for the engine. Unlike in the world of cargo shiping, our time was our own.

CURRENT

Wandering the dock in that microseason of waiting, I met Susan Robertson Rooms, a seventy-four-year-old sailor in an adjacent vessel. Seeing my cardboard sign, she invited me on the deck of her monohull to chat. "Tell me a story about water? What's that all about?"

She told me that she had owned this boat—the *Assagi*—for forty years. "My dad and I built it in Melbourne," she said.

Each of her stories had a sliver of danger nestled in its core. Everything that could have gone wrong on the water did, and yet she was still there—a testament to her own survival.

Susan led with a story of a wave in Bass Strait near the shipping entrance to Melbourne that rolled her boat completely. "Everything went wrong, but no one was hurt," she said. "I thought I'd relate something a little bit more pleasant."

In 1976, Susan sailed to Tahiti from Australia. The first leg crossed the Tasman Sea. "We went like blazes," she said. "We had big seas, following seas." The waves were moving in the direction of their travel, speeding along the journey.

After spending a year in New Zealand, Susan's friend from Mauritius told her that she had to sail to Tahiti. "You crossed the Tasman. The rest is easy."

"He wasn't quite right about that," Susan said, not elaborating on the negative. When she arrived, Susan entered a race around Tahiti—176 nautical miles. If you start your engine, you're disqualified. "The whole fleet stops halfway and the little town of Taravao shuts down, and there's a big party for the time that the yachts are there.

"What will never be removed from my memory is the dawn coming back around between Tahiti and Mooréa, which is a place in the old days where ships without engines used to get caught in the current and they would crunch the reef and that's the end of them." At night, she told me, a cold land breeze came down from the mountain. The land breeze blew until a bit after dawn.

The tactic, in order to avoid being becalmed between Tahiti and Mooréa—a space where you don't have any breeze except for the land breeze as the sailors were on the lee side of the big island—was to rely on sound. "If you can't hear the surf crashing on the reef, you're too far out and you probably get becalmed, so you sail closer to the reef in order to get to be able to hear it," she said.

Being attuned to all the senses while traveling on water is crucial for both movement and survival.

"But the trouble is, if you go too close—and remember, we're talking in days where there was no GPS and none of this business of working out exactly where you are to the nearest few meters—if you go too close, you could get sucked in by a big wave and smashed onto the reef," Susan said.

The second cue was visual. At dawn, after the sun crested the mountain, the turquoise parts of the water denoted a sandy bottom. "As the waves stood up just before it broke on the reef, you could see through it and what you saw was the color of the sun reflecting off the bottom and through to where we were. And of course, that was okay for the sailing after that, because you could see the bloody reef," Susan said. "That was one of the things that really stuck in my mind."

This was the 1970s, though. Things have changed. "From what I can hear talking to other yachties who've been to Polynesia more recently than I, the colors are still the same, basically," Susan said. "I don't know how ocean warming has affected the coral, but I know from reports it's monitored very closely by scientists," she said, noting that where we were sitting, at the port in Townsville, was near the Australian Institute of Marine Science and researchers at James Cook University. "They know the ocean warming gets to a certain point and the corals are going to die and bleach," she said. "I don't know what it will be like in Tahiti, but hopefully I'll get back there. It's a long way to windward, though, for someone my age." Susan laughed. A new generation of yachties would have to be her eyes.

After the race, a Tahitian invited her to go spear fishing. "I remember diving and being absolutely overwhelmed by the beautiful color of all the corals." And there, with her snorkel mask, four miles away from the island, Susan decided to go for a swim.

"Big mistake," she said, laughing grimly.

"I was lucky I had a full-on wetsuit. Because I thought, *Oh, I'll go and see what it's like deeper*. And I was in a gutter and the surf's breaking from outside and every time a wave came in, I hung on to the coral. And when the surge had gone, I swam with it going back again. So this happened three times. The third time I went, there was no coral under me." She barely managed to scurry her way back to the boat before being swept out to sea.

Even a strong woman of the ocean is still vulnerable to its whims. Water both creates and destroys, but as this sailor's stories show, there's no reward without a taste of risk.

CHURN

At last, the gasket arrived. We left the Townsville Marina heading north toward Thursday Island, motoring at first, then swept up by the wind. Jenny lit up on the water: this is home. As a new crew member, I did my best to learn how to harness the wind. Mostly, though, I washed the dishes. I was delegated once to making the coffee, but Randy crinkled his nose and deemed it too weak.

Once we left the shelter of land, the waves got stronger. My stomach started to turn. This was nothing like being on board a cargo ship, where the waves propagated long and lean through the steel. No, we positively bobbed through it in more than four dimensions, a rotating log going up and over a waterfall, then down again, then sideways. My stomach rebelled. This sea goddess in training puked three times in the first eighteen hours. Deflated, I could barely keep down a saltine cracker. Jenny coaxed me to drink more water and stroked my head until I fell asleep. Being at sea is an act of surrender.

I learned to laugh at myself and my stomach. I learned that squalls give us speed (and rain!). I wasn't able to do much aside from sleep and be horizontal. Everything made me nauseous. In between his navigation shifts, Bruce looked at me skeptically. I closed my eyes and tried to apply some kind of order to the bouncing horizon. After twenty-four hours, I could keep down saltines, saltines with a bit of butter, and a tiny piece of bread with melted cheese on top.

CATCH

Everything changed with a fish. Bruce cast a line off the stern and caught a blue mackerel longer than my torso. Randy filleted the fish on the deck, and Jenny marinated its flesh in soy sauce, garlic, pickled ginger, and sesame oil—the strength I would soon welcome into my own. The muscle of it.

Then, moored on the coast off Thursday Island, Captain Bruce asked me to leave. "We can't keep taking you," he said. "It's too risky."

JUMPING SHIP

In the morning, as fate would have it, another boat appeared—a nineteen-meter catamaran called the *SV Pelican*. They must have arrived in darkness, moored in the same spot for the night. I suggested to Jenny that we share our fish. Without refrigeration on board, we had more than we could eat before it spoiled in the heat. Jenny and I paddled over on an inflatable raft with the spare filets. Their captain, Garry McKechnie, told us that the *Pelican* was set up for marine research, environmental education, and community engagement. In the past they had studied seagrass and coral bleaching. They were returning home to Bermagui, New South Wales, from Cape York in far North Queensland, where they had been working with the Hope Vale Indigenous community.

Garry pivoted to talking about airplanes. "I think that there's a cultural blind spot with regard to flying," he told me. "When I was a very young person, we had to save up for years to be able to afford to fly to Europe. Well, the airfares now are cheaper in straight dollar terms than they were when I was nineteen, which is forty years ago." When the captain was nineteen, the trip cost him two thousand dollars to fly return from Australia to London. In 2015, that same trip would cost as little as $1,349—and that's not accounting for inflation.[6] "It shouldn't be that cheap!" Garry said.

"People love their international travel," Garry continued, "but it's not very fashionable to talk about what that is costing us in environmental terms. I'd like to see more people becoming a little bit more aware." In addition to being skeptical of flying, he had his doubts about the benefits of tourism. "Overall, I don't know that tourism is actually good for anyone, which includes the people who are touring and the people who are being toured to, if you know what I mean," he added. "We're going down a road which doesn't have a happy ending for anyone. And certainly not for the planet."

For Garry, the loss of biodiversity—the sixth great extinction event, and the only one created by humans—was the most concerning. "I see no sign that we're actually dealing with that as a species or as a culture," he said. "The reality is that overall, the human species is overrunning this planet. And we're kind of taking it down. Sorry that's a bit of a downer," he laughed. The water lapped gently at the side of the boat.

"I'm curious. What do you like most about being on the water?" I asked.

"I grew up by the water; I grew up around boats," he replied. "My parents used to take me camping to amazing, remote places. In very small boats, we would go out to islands that no one else went to."

Garry's first career was in the arts, but he always wanted to travel and work on a boat. In his midforties, he made that dream happen. "That's what I've been doing the last twenty or so years," he said. "I love the ocean. I love the water. I love the outdoors."

After our interview, Garry mentioned that he was looking for a backpacker to help them with the night shift, sailing south.

I volunteered myself. He said yes. I packed up my bicycle and jumped ship. Jenny helped me carry my panniers over the water in an inflatable dinghy.

PELICAN

So this is the plot twist: I joined ranks with a cook, a piano player, and a trumpeter. As a goodbye, Jenny gave me a seashell anklet. I gave her a piece of Whanganui river glass. Ships are contained social spaces. I exited one and entered another, not knowing what to expect. The mackerel provided the pivot point between following the route I thought I was on and surrendering to the flow of going where I needed to be. The ocean brought this about. I jumped ship, moving backward and forward at once.

The first thing to note: catamarans are more stable. My stomach was, blissfully, at ease. Garry caught two more fish: a mackerel and a tuna—both impossibly large. I watched their scales as they died—the captain's unbridled joy: slicing and barbecuing and eating. The color and texture of the fish skin looked like every color of the sea as it changed throughout the day: gray to pinkish to dusky blue. In the evenings, the crew made music—Garry on the keyboard, and Andy, the galley cook, on the trumpet. Sometimes one or the other of them would belt out a song—louder than the waves, louder than the wind. I loved being so infinitesimal and singing in the face of it.

On the mesh of the bow, Andy taught me how to jump over waves. Later, as we washed dishes together, he anticipated me asking if I should sleep with my head toward bow or to stern. "Stern is best," he said, elbows deep in the suds.

The engine ran all night and into the next day to power us through unfavorable winds. The Pacific cradled me. I learned the sound of the rain as it slapped and hit the sea.

Andy's first lesson: "Do you trust the boat?"

Even during some loud, rocky, bumpy nights, I trusted the boat. I was not afraid. Or rather, my fear didn't outsize me. I brought that fear to the dining room table to talk.

NIGHT SHIFT

The waves had their own tongues: wild horses or wolves. The sun slipped below the surface. At this moment, aboard the *Pelican*, we floated into a loud boom kind of storm. I bounced about my cabin: a wild night, overseen by the soft sliver of a moon. The ocean was rough. I couldn't sleep. Slow travel has its perils.

For a few days, the Great Barrier Reef sheltered the catamaran as we sailed south. Now, the boat was irreconcilably bumpy. After dinner, I stole a few hours of rest before night watch began. My ears were full of the humming engine and the water.

Each shift lasted for four hours: Hans, a quiet, white-haired man, kept watch from 7 p.m. to 11 p.m., I took over the 11 p.m. to 3 a.m. shift, and Andy rounded off the night by watching from 3 a.m. to 7 a.m., at which point Captain Garry woke up to squint and oversee the day.

During my shift, I shook awake Captain Garry if the wind direction changed dramatically or if there were any approaching lights on the horizon. Otherwise, I was on my own, monitoring, bearing witness. I loved the dark silence of the stars.

We sailed the reverse of the trip I made up Australia's east coast on my bicycle. I remembered snippets as the land slid by: Lady Musgrave Island, Seventeen Seventy, Agnes Water, scrunching my toes in the off-white sand. In the daytime I cleaned the galley and caught up on sleep.

I was grateful for the way the surface of the sea reflected the sun's many moods. I was grateful for the stumble ballet of being on board a sailboat and the reminder to be light-footed that water provides. I was grateful for this zigzag journey, for a year's worth of conversations that changed and moved and showed me that great people are everywhere, if only I stay open to meeting them.

The waves pitched and rolled. The wings of the sails lifted into the golden hour. I practiced tying knots: a clove hitch, a bow line, the figure eight. The sails were tall ghosts. To my right, in the wheelhouse, Captain Garry snoozed deeply on the bench, his mouth open. I listened to his breath, the waves under the hull. Our navigation system glowed green. Sometimes I wandered on the deck. Garry told me not to fall—that I would die. The rest of the crew was asleep. No one would know.

WIND

This much I know about the wind: it can (and will) change. A gust. I shook Captain Garry awake to check that we hadn't veered too far out to sea. We could see the far-off lights of Brisbane on the starboard side. Quite a bit of light—hazy and sprawling and low. What do I remember about Brisbane? The bike paths. My Warmshowers hosts. The ferry. If I didn't know better, I might have thought that the Brisbane lights could be the sunrise. My bicycle was stashed safely under the captain's bed in the wheelhouse. I didn't make it to Indonesia, yes, but this change of course was just fine.

LIGHT

One night, I scanned the horizon on night watch. Everyone else was asleep. The captain snored beside me in the wheelhouse. I knew my instructions: if the wind shifted more than five degrees, if we started to drift too far off course, or if the light of a cargo ship seemed to be intersecting with our path, to shake the captain awake.

Without warning, a large, circular light appeared. I thought it was a cargo ship, barreling down at us. Panicking, I shook awake Captain Garry.

"We're about to get run over!" I told him.

"Devi," he said. "You just woke me up for the moon."

MUSIC

Andy, the cook from New South Wales, attended to the connective tissues in interpersonal relationships, so vital on a boat. After meals, we retreated to the galley to talk, washing dishes with Dr. Bronner's soap.

Our hands full of suds, Andy told me about his dead dad's trombone. He picked up music after his dad passed away so that he could be closer to his father in death. Andy is number eleven of sixteen brothers and sisters. His dad was a butcher and a trombone player who died unexpectedly in his sleep. His jazz band was called Zenith.

The boat was our classroom. I taught Andy how to fold paper cranes and how to do a time step in tap dancing. He always wanted to learn tap but never did, though two of his sisters took lessons for a bit and taught him a move or two.

We stopped for the night at Fraser Island, so peaceful to have silence from the engines. After dusk, Andy and I went out to the bow to look up. The lights on the masts looked like constellations. There were more shooting stars than I could count. I wished for a safe passage. "Lead by the most delicate spot," Andy said.

In the morning, we sailed by the power of the wind alone. Sometimes we had visitors: the water was so clear that I could see straight to the sandy bottom. Two dolphins made a mammalian sound as they surfaced—breathing—to surf under the bow. I lay on top of the mesh there, watching them, whooping for joy.

Another night, just after sunset, I saw what I thought was a big buoy rise up out of the water, far off—but it wasn't a buoy. It was a whale! A humpback twisted up and fell down onto the sea with a splash. I shouted and rushed to ask Andy and Garry if they had seen the whale breaching, too.

They hadn't. But they shared in my joy for a moment. The space of the waves where the whale's body had slapped down was miniature eruption. A deep breath.

LAND

The sunrise opened. Land returned to the horizon: the center rather than the periphery. My center of gravity shifted. At sea, I was someone else. I had fewer borders. I was less afraid to experiment. I caught myself when I fell down. By month's end I spent more of October 2015 on the water than on land: monohull to catamaran to cargo ship—*Far Fetched* to *Pelican* to *Spirit of Shanghai.* Captain Garry saw wolves in the waves, whales he was certain were not there. I could tell it was almost time to leave.

"You can anticipate someone's future actions by looking at their breath—a technique of improvisation," Andy told me, a parting gift. "Don't forget to play," he added. "And to go swimming."

The hardest part of this goodbye was knowing that this crew was temporary. They welcomed me and, in all likelihood, we will never see each other again. I knew I would miss the ease of this water family. I would miss being barefoot.

I put my shoes on. The land felt strange, too solid. I bent my knees when I moved in a way that was incompatible with the texture of concrete and soil. My crewmates, off to see their land families, waved goodbye. Our tight social unit dispersed. We became strangers again.

In Bermagui, I took a few moments to recoup alone on a picnic bench and sip a cappuccino—the small comforts of a continent—before taking the bus and train up to Sydney, where I met the cargo ship that took me back to New Zealand. There are no straight lines.

6

THAILAND

SPIRIT SHRINE

The year of no flying ended like this: I gave up. After four months of trying unsuccessfully to secure a passage on a sailboat from New Zealand to anywhere-that-was-not-Australia, I bought a cheap flight to Bangkok in May 2016. I packed my bicycle in a cardboard box. I flew. It seemed, at the time, to be the only way forward. It was a strange feeling, being on a plane after having avoided air travel for so long. My feet hurt. The air was so dry. Cargo ships are loud, but airplanes are louder.

Two flights, an eight-hour delay, and a taxi ride later, I caught the sunrise in my open arms. A friend from my hometown let me stay at her sixteenth-floor flat in Sukhumvit. I ate a dragon fruit on her balcony, above the chaos of power lines, traffic, and trees. Some trees, decked out in lengths of colorful cloth, are protected. I melted in the heat. Nearly every building, I learned, has a spirit shrine. People bow at the spirit shrines as they pass.

Sema Thai Marionette is a puppetry company in Bangkok dedicated to preserving and performing Thai puppetry traditions. I went to their morning performance of a show about the Rambutan prince at a primary school, and then hung out at their puppetry workshop for the afternoon. The team of puppeteers helped me translate my cardboard sign into Thai so that I could collect stories on the street. They also let me try out a bicycling marionette—the pedals moved with a flick of my wrist.

MIGRATION

Water is movement. Mess with the amount of water available in one location, and people will move if they must; we all need a steady and semi-predictable supply of water in order to survive.

Sun is a modern dancer who moved to Bangkok from the rural north. He relocated to the city in part to practice his art, but also to take refuge from unpredictable rain patterns. Farming in Thailand is governed by the monsoon season, a shift in the direction of prevailing winds that dump rain, fill river basins, and irrigate crops from roughly May to September. Or at least they used to.

When we spoke in late May 2016, it was dry in Thailand. The seasonal rains were delayed. Water levels in the country's biggest dams plummeted to less than 10 percent of their capacity—the worst drought in two decades.[1]

"Right now it's supposed to be the beginning of the rainy season, but there is no rain," Sun told me. "How can I say it? I think the balance of the weather is changing. Some parts have a lot of rain, but some parts have none." He leaned back in his chair, expressing the dryness and imbalance with his hands moving like a fulcrum scale.

"That is the problem. The people who used to be farmers have to come to Bangkok because they want money and they want work," he said. "There is no more work because of the weather."

Migration to the city, in other words, is hastened by the rain.

POLLUTION

Pollution was on the tip of many people's tongues. Asked to speak about water, it was the first thing that came to mind. In Bangkok, a woman spoke to me in front of her home on Lat Phrao Road, in front of Khlong Saen Saep, a canal. "The canal was not this smelly in the past. But now, even a drop of water splashing when a boat passes by smells so bad," she said.

She told me that the day before she had overheard two men talking about this. One of them said, "Do you want to go back home by boat?"

"No, I don't want to," he replied. "Yesterday, water splashed on me and it smelled awful."

"I was so embarrassed," the woman told me. "In the past it wasn't like this. I lived here for twenty years. The smell kept getting worse. The canal is long. The canal around Min Buri district does not smell bad. I do not know why it smells bad here, in the city. I have been here for more than twenty years, but the smell is still bad. Please help me."

In Ayutthaya, a high school student, Wanmingkwan Kumnerdrat, told me that this city, previously Thailand's capital, is surrounded by three rivers. One of them, the Chao Phraya, has been used for centuries as a source of fish, water, drainage, and recreation. "The water condition is not as good in Thailand as it was in the past, because in the past the water was clear," she told me. "In Thai history they say that we could see the fish in the water. But now we cannot see them because of all the water pollution."

She didn't specify which kind of pollution.

The next day, at a food market just outside town, a vendor stopped sweeping to tell me her story. "There are so many factories around here. They dump chemical substances in the river and kill all the fish," she said. "So many fishes died when they dumped their waste in the Pa Sak River. They dumped it, but they did not clean up their mess. It's not clean, and it's not safe," she added.

Over time, the toxins accumulated.

"I saw so many dead fish. There was some kind of foam, too. Then the fish were dead. There was chemical powder in it," she explained, leaning against her broom. "People around the river contacted the government so many times, but it's still the same," she continued, shaking her head. "Those factories do not clean the river for us, but they keep dumping their waste. Then, the aquatic life is dead. For humans, it is not drinkable."

On the road to Chaiyaphum in Thep Sathit, a district in the southwestern part of the province, a third storyteller, a woman in her fifties, spoke to me about contamination. "In the past, the water was so clear and clean that I could drink it, but now, I wouldn't try. It has been contaminated by herbicides and other chemical substances," she said. "When I was a child, I saw Thep Prathan Waterfall with my friends. The water was crystal clear. It's so different now. The water is opaque. If it could be just like before, it would be great." Now, she told me, "swimming in the river would burn your skin."

The expansion of agriculture is in part to blame. "We had more forests in the past, and the lands weren't used for agriculture that much. As the time passed, more people cultivated their lands, and when the rain poured down, it would push the soil in the river, too. It was contaminated by chemical fertilizer," she told me.

The rainwater used to be drinkable, but now, "we have to drink from a bottle instead because the rain smells weird due to gas and dust. Sometimes, I find some black sediment in rainwater. The rainwater used to be clean and clear, but now, I have to buy clean water."

PICKLED FISH

Bangkok left me tense and flighty. I took a train to get out of the worst snarls of the city, and pushed through the backroads, the edges, the agricultural areas, pedaling northeast toward Chaiyaphum. At dawn, monks in orange robes walked single file, bowls in their hands, receiving morning alms of rice or fruit. Outside one temple, a woman blessed a piece of white string and tied it around my wrist. Cows, which I initially mistook for water buffalo, raised their heads as I passed, bells tinkling on their necks. I passed a tree with a shrine at its base—cloth wrapped around the trunk, lush in its colors: an omen of what was to come. I didn't always have the vocabulary to describe what I saw around me. Pedaling through this landscape was an experience of interiority. I dodged puddles and negotiated the sparse, rural traffic. I coughed in the exhaust. Most of the time, I was alone with my thoughts.

By midmorning, in the climbing heat, I made a wrong turn on a levee, and followed its elevated spine around the edge of a lake. Seated there, on the banks,

I found two storytellers, both farmers. I pointed to my ear to signal listening, then mimed the use of a microphone. They said yes. I listened. Years later, I gave a talk at a high school in New Hampshire, and Warich Ngamkanjanarat, an exchange student from Thailand, offered to translate what one man had said:

"My wife and I came here by a farm tractor to catch fish using a fish trap. We use these fish to make Thai-style pickled fish, which we use as a substitute for salt," the man told me. "We put it in soup, spicy soup, or eat it with rice. We made it by mixing salt-pickled fish with yellow ground roasted rice. Most of the time we use local fish that can be used to cook something else, too, such as a spicy dip."

After the story-recipe, the couple pointed me on the right route toward Chaiyaphum. Even without a common language, gesturing in one direction and asking "Chaiyaphum?" was enough to get by.

PHI FAH

On the narrow dirt road, surrounded by the water of rice fields, I rounded a corner to find a dozen people dancing around a tree. Two women sang and two men played tall instruments made of bamboo. One of the dancers flagged me down to join. I left my bicycle on its side; they wrapped a red scarf around my neck and a length of red fabric around my hips. We shuffled our feet and kept our arms loose from the elbows down, wrists cycling, to welcome the spirits. Others moved inward to offer incense and bottles of Coca-Cola and yellow Fanta to a shrine at our feet. At the center of the offerings, a mirror on the ground reflected the tree's branches back at itself. A dozen feet away, a spirit house sat at the edge of the field, small but mighty, incense burning out the front door.

Time passed—I'm not sure how much. I entered a fugue-like trance. There was nothing but the tree, nothing but the music. My movement became almost involuntary. I faded to the outside of the circle, planning to get back on the bicycle, to leave the tree. But the woman who had ushered me into the ceremony called me back to eat, picnic-blanket-style, under the shade. The musicians put down their instruments. We shared plates of rice and fruit.

"It is pretty fertile around my home," one of the dancers told me, her hand full of sticky rice. The Chi River provides water to irrigate the crops. "We don't have much money, but we want everyone to be able to afford their basic needs. We live sufficiently." A steady supply of rain is critical to this existence, and the ceremony helps keep it that way. She danced to honor a holy spirit, Phi Fah, who provides rain, can cure people, and gets rid of bad luck. Each year, a guardian of the village tells a fortune predicting whether enough rain will fall. "If not, we have to perform the *Rum Phi Fah*," she said.

A second option is to light a ceremonial rocket called *Bang Fai*. "If the rocket flies higher, it means that it will rain a lot. If the rocket flies lower or explodes, it means that there will tend to be a drought that year," she told me.

Another dancer chimed in. "Our ancestors from a long time ago passed down this tradition again and again. Parents passed down the tradition to the next generations before they died. It is the tradition of our province and other provinces in Isan, northeastern Thailand," she said. This ceremonial dance happens between June and August, before the Buddhist Lent. When the Buddhist Lent starts, there will be no more *Rum Phi Fah* until the sixth month of the next year. The dancer paused to stare at the three's thick trunk before continuing.

"We do this annually until we die. I do not know when the tradition will come to an end," she added, her voice trailing off. "We will see." She sounded like she knew that one day, the dancing would stop.

After feasting on rice, we danced again to feed the spirit. After another several rotations around the tree, I managed to leave, my belly full, my arms sunburned. It felt like dreaming. I cycled into another dimension. I rounded a bend and couldn't hear the instruments anymore. With the unpredictable rains, how many of those dancers would move to Bangkok? Which traditions will survive? Which will wash away? Generations from now, who will feed Phi Fah? Who will play the bamboo instrument, dance in a circle, appease the hungry spirit of the rain?

CHAIYAPHUM CYCLING CLUB

Bike touring is an act of repetition. Most days are similar—wake up, push, balance, listen, float. In New Zealand, I developed a ritual of stopping at swing sets. Every time I saw a playground, I would get off the bike and pump my legs into the air, disassociating into some fragment of the divine. In Thailand, there were temples, a useful punctuation mark to hours in the saddle. Every time I saw one, I swung my leg over the top tube and unbuckled my helmet to pray. I knew that I was probably doing it wrong, but it helped to have a routine. When every day was constant motion, little things helped keep me grounded.

Within a few kilometers of Chaiyaphum, a small city of nearly sixty thousand in Thailand's northeast, members of the local cycling club met me at a temple. Legs heavy with the day's miles, I drafted behind a woman with a long braid, talking about our favorite times of day. Mine is the golden hour, when the sun is ripe and setting. Hers is eight a.m., before the day's heat opens. We raced the sun.

That night, home was a shack near the rice paddies. I dropped my panniers and headed to the night market to eat mushrooms and rice and papaya *bok bok* salad. Everything tastes better after spending all day outside. Over coconut ice cream, the cyclists told me about their favorite routes. Yongyut Choomee prefers to tour alone, or sometimes with his wife. Cycling helped him unwind from his work as a contractor. "Others might say it is hard, it is tiring, and it is torturing, but when we reach a destination, we always see good things. Sometimes, I meet people. Sometimes I don't," he told me. "Biking gives us inspiration to do something."

Yongyut told me about his tours. In northeastern Thailand, near Phu Chong Na Yoi National Park in Ubon Ratchathani, thousands of shrimp, each the size of a thumbnail, walk out of their streams at night during the rainy season.[2] This part of their upstream migration helps them detour around strong currents; the shrimp march on land and return to an area where the water is less turbulent. As they march, the shrimp travel in a continuous parade. Individual shrimp can walk for distances of up to sixty-five feet. Frogs, lizards, snakes, and spiders all prey on them when they leave the water. Tourists shine flashlights onto the riverbanks to watch the creatures move.

His other favorite destination, Bak Teo Waterfall, was named for a hunter-gatherer who came to the area to collect beehives. The hunter used a vine rope to jump out of the waterfall, but in midair he realized that he was holding on to a sacred snake. Frightened, he used his knife to cut the snake and fell from the cliff.

"Villagers believe this happened because Bak Teo came here to collect forest products without asking permission from the guardian of the forest," Yongyut told me.

TULIPS

Another member of the cycling club, Tom Jai, told me that she had recently had a conversation with students attending secondary school in Roi Et province. "We discussed how the low water level in a dam makes people not have enough water," she said.

One student answered, "Why don't they just use tap water?"

"This shocked me," Tom Jai said. "It makes me sad that our new generation thinks like this." To her, water management and climate change are intertwined; to not prepare children for this reality is a disservice to their future. "This made me realize that their understanding of water management and global warming is really poor. It shows that they do not care about the problem at all," she added. Education is key.

Tom Jai also wanted to tell me about Siam tulips: pink, spiky, and native to this part of Thailand. "Global warming is one of the factors that has affected the exporting of Siam tulips for the past two or three years. It caused delayed rain, which affected the tourism in the end," she said.

"Because Siam tulips only bloom for three months, plus the global warming, it sometimes disappointed tourists who expected to see the tulips bloom during a specific time."

For Tom Jai, life and water are intertwined. "I want everyone to realize that we have the forest, we have water, thus we have life. If we don't have the forest or water, it will be hard to live. This is also a saying of the Rama IX king and queen. If we care about the environment, we will be able to live in this world happily," she added. "Living with something too artificial makes you ill."

CLOUD SEEDING

I met Sathon Wong-Od, an airplane mechanic, in a monsoon. His breath was heavy with whiskey. Around us, rain fell, relentless. Sathon spoke with pride of Thailand's cloud seeding program, led by the Department of Royal Rainmaking and Agricultural Aviation.

"They tell us how high and where we need to go. When they confirm it, we'll begin the cloud seeding process there," Sathon told me.

The technology underpinning these efforts to change the weather dates to the 1940s. In 1946, Vincent Schaefer and Bernard Vonnegut of General Electric seeded a cloud floating above Schenectady, New York, with dry ice. Snow fell from its base. Eventually they settled on silver iodide, which has crystals structurally similar to ice, as a way of inducing precipitation in a cloud by promoting condensation. Silver iodide can be sprayed on clouds from an airplane or sent up from a rocket.[3]

"The academic department loads these substances on a plane. When the plane is at the right altitude, they dump the substances out of the plane," he said. Then, the rain falls. Sometimes, it takes a few hours for the condensation to work. But only a tiny bit of the chemical is required to make a difference. Sathon drew a diagram in my notebook—the plane, the cloud, the droplets.

"Can you now understand?" he said.

Thailand is not alone in this effort; more than fifty countries have active cloud seeding programs.[4] I had to silence my audio recorder to protect it from the rain dripping off the edge of the roof.

CONTROL

And then, of course, there is the impermanence of rain—its ability to resist even the best-planned efforts at control.

"My home is in a rural area. The temperature is lower than in Bangkok. During the rainy season, we get water from nature," one storyteller told me. "We can't control it. It's rainwater."

Sometimes, there is poetry in simplicity.

7
LAOS

RED MUD

On a bicycle, there are no windshield wipers. When it rained, I tried to find shelter. If I couldn't, I got wet.

Past the Chong Mek crossing point from Thailand, I detoured from the sealed road. Riding through heavy mud is slow going. My tires pressed into the earth, carrying cakes of red into the derailleur and splattering dirt up the back of my legs.

I stopped to buy bottled water at a roadside stall. The man at the counter told me it was the beginning of rice planting season. "Rainfall is seasonal, but it's not as good as it should be. In the rainy season, it feels cool and wet, and in the dry season, the weather is very hot," he said, matter-of-fact. "There may be limited rainfall in some areas due to climate change caused by deforestation."

He handed me my bottle of water and I pedaled away.

LUCKY

There's a tradition of hospitality in Buddhist temples—if a traveler asks to stay, the monks will almost always say yes.

In Pakse, a city on the banks of the Mekong, I navigated toward Wat Luang, a temple with tiered roofs and carved wooden doors. Butterflies punctuated the sky, tiny against the towering rain clouds. Temples are easy to find; their roofs are the pointiest in town. After getting approval from the higher-ups, a group of novice monks in orange robes directed me to pitch my tent under an awning by the river. When the clouds spilled over, a family of kittens bumped against the edge of my tent. We were dry. This was a blessing. The lightning glowed purple.

In the morning, Anouluck, one of the teenage monks studying at the temple, translated my cardboard sign into Lao. He told me to call him Lucky for short. A group of his friends gathered around a table to hear his water story. Lucky told us that he comes from the Salavan province, just north of here, and had been studying at Wat Luang for two years. More than anything else, more than practicing Buddhism, even, he loves to play the guitar. Lucky told me that his mother sent him to study at the temple because it was the most affordable option.

A river flows through Lucky's village. He traced its path with his fingers on the table: it flows downhill to Pakse, where it meets the Mekong. "There are crabs, fish, shrimp, and snails," Lucky said. "People live on the river as a source of livelihood."

GIANT

Over a thousand years ago, Khmer people built the Vat Phou temple complex at the base of a mountain in southern Laos, in honor of the god Shiva. The temple was recognized as a UNESCO World Heritage Site in 2001.[1] One evening in June 2016, a man named Bouasone told me that a spring at the top of the mountain provided holy water—"spirit water," he called it—to the worshippers. Health, luck, or strength would befall anyone who bathed in it.

The spirit water, according to legend, could also grant immunity from death. Shiva drank it. One day in the distant past, a human-eating giant arrived at Vat Phou. He had heard about the spirit water and he wanted a taste. He hung from the sky and jumped on the mountain and used his big finger to attract the water. "He drank it very quickly," Bouasone said. "But some people told him, 'You are bad. You cannot win. This water is spirit water. Only the best people can drink it, like Shiva.'"

The giant, defiant, said, "I can drink it! I don't want to die."

While he was drinking, the worshippers tried to cut the giant's neck. But the water, by this time, had reached the giant's mouth. His lower body died, but his head stayed alive, immortal. The water preserved him. The giant continued to try to eat people with his big mouth and his big teeth.

Then Shiva arrived. He saw that the giant's head would never die, but he wanted him to stop eating people. The giant promised, but Shiva knew it was a lie. "I don't believe that," Shiva said. "I have to sit on your head forever."

Bouasone pointed to a picture of a statue of Shiva and the giant in the ruins of Vat Phou. "Shiva sat on his head to control him," he said. "But he started to eat people again."

A peal of distant thunder echoed through the room.

MEKONG

Rain is the rhythm by which everything else happens. The evening came on like it always does. The bird-flying time, the cooler air, the many creatures locating a place to rest. Sometimes, a little wind comes up. The light is magnificent, gold. The mighty Mekong, brown with sediment. The broad back of the river and the boats that cross it. The puddles, the fish. The swaths of misty green.

At another temple, I slept next to two gongs. In the morning, someone rang them. I continued to sleep. When I woke up, thick rain was falling into the Mekong. To pass the time, the monks sat under an open-air hut along the river, playing music through a speaker and catching the breeze. Under the platform, a white rabbit gnawed at its pink cage next to a monkey with three legs, chained around his waist to a beam of wood supporting the platform. Above, some of the senior monks slept on their backs.

The big droplets lightened, then stopped. Two roosters crowed: one close, one far. The river shone. The monks put their orange umbrellas away. A man in a white helmet drove his motorcycle through the temple grounds, leaving a line in the mud. A pack of five chickens roamed the temple grounds, pecking for worms. Later, during prayer, the monks chanted in uneven rhythms, then bowed their heads to meet the floor.

8
CAMBODIA

BREATH

I only boarded the van in the first place because it was raining. The monsoons had arrived. The roads turned to mud. My tent leaked—fat drops that pressed through the seam above my head at night, keeping me awake and shivering. The thunder was the loudest. Big wind rattled my tent poles. I wasn't ready to keep biking, not like this.

So I bought a ticket that would take me across the border to Phnom Penh. On that journey, two men (one from Laos, the other American) told me that they would kill me. When I tried to take a photo of their faces, the Laotian man mimed slitting my throat and tried to shut the van door on my body. A few hours later, when we arrived in Phnom Penh, the American man pressed his hands around my neck and started to strangle me. A tuk tuk driver in Phnom Penh pushed him off.

At some point during the fourteen-hour ride, I told the American man sitting next to me that he was "mansplaining"—a term popularized by writer Rebecca Solnit.[1] Mansplaining, for those unfamiliar with the term, happens when men explain things to women that they already know. This phenomenon, unsurprisingly, knows no boundaries or borders.

In the van in Laos, I mentioned that it was weird that we were stopping for half an hour outside an abandoned warehouse, and that an unknown person added a cardboard box full of something to our back seat and walked away. We drove off without that passenger. It seemed like the driver was using tourists as a cover to smuggle drugs across the border.

The American guy said, “If you had traveled more, you would know that this is completely normal.”

I said, “Dude, I’m telling this to you because I value you as a human being, but there’s this term called mansplaining where men explain things to women that they already know—”

“You’re one of those fucking feminists,” he cut me off.

Without thinking, I responded. “Yes, I am.”

We then engaged in a war of manspreading in the back seat. The American was trying to take up as much space as was physically possible by spreading his legs wide, so I did the same. He did not respond well. He started commenting on my crotch and my body hair, insinuating that I must have a penis.

People kept adding things to the van at each stop, some very heavy, and I was worried that my bike frame, which was disassembled and stashed under the back seat, would crack under the weight. At every stop I would get out to check on my bike. The Laotian driver, fed up, threw a Coke can at my head. I ducked.

A few hours later the American called me a “feminazi cunt.” He said that he hoped that the driver would beat me up so that he could watch, smoke a cigarette, and laugh. He said that I deserved it. When I asked him why, he admitted that he didn’t see me as human. “Women like you are the reason why women get beat up,” he said.

There were other people on the vehicle: mostly tourists and a few locals. They didn’t intervene. This is why I need feminism. Gender-based violence happens. At the root of a lot of it, I think, is the idea that women are not human, or are somehow less human than men. Margaret Atwood once wrote: “Men are afraid that women will laugh at them. Women are afraid that men will kill them.” I was afraid. If I made a mistake, it was in engaging with the man. But given that we were crammed next to each other for fourteen hours in the back seat of a ten-passenger van, I didn’t have much choice.

After that violent day, I had panic attacks. I didn’t know the word for it then, but my whole body felt like it was shutting down. I couldn’t breathe. I didn’t know where I was. The smallest things would set me off: downed power lines, chaotic traffic, gathering monsoon clouds, potholes in the sidewalk, the slap of dust in my face before a storm. Even a stranger making direct eye contact was too much. A woman in the van later told me that it was my fault, that I had provoked him.

I disassociated. I was a child again. Four years old. Corvallis, Oregon. In a different rainstorm, I watched the ones I loved get hurt—lamps thrown across the room, my mother held up against the wall by her neck—but I couldn't do anything about it. I was powerless to protect those I loved from those who were supposed to protect me. The fabric of time ripped, to the place in my life when I had been the most afraid. This scene of domestic violence, long gone, roared in my head. Such is the knife of panic. The scariest part was not knowing where I was, or when. Every physical journey, I suppose, is also a journey into the recesses of one's own mind. From the outside, it looked like I was sobbing. Inside, I had forgotten how to breathe.

FIRE

Days later, doing my best to regain a sense of calm, I met a Cambodian cyclist, Rithy Thul, in Phnom Penh. In 2009, he cycled from his home in Siem Reap to Kulen Mountain, a place believed to have a high concentration of spirits. It was the rainy season, and Rithy needed to get out of the city. He took his tent and bicycle and pedaled up to spend a week in the mountains. Along the way, he passed small villages, one after another.

"There's nothing to sell, just people who say hello." Some had machetes on their shoulders. "They look like they could be murderers, but they are friendly," he said. "They just do their work, and they want to talk to me because I look different. I'm from the city, and they want to know what the city's like. Some of them have visited the city but have never lived there."

After a long day in the saddle, Rithy was looking for somewhere to sleep. He found a pagoda, a concrete building where monks lived and cooked and prayed. He asked the head monk for a place to stay the night. The monk told him to put his tent or hammock near a building used for gatherings. "It looked like it was almost going to fall down, so I didn't want to sleep there," Rithy told me. He asked the monk if it was okay to sleep on the mountain, on top of the hill.

And the monk said, "Yes, but you think you want to sleep there?"

Rithy thought to himself, *I'm pretty bold.* He said yes. The monk showed Rithy his own spot for meditating, away from people. The monk said, "You can sleep up here, but only on this open space of rocks." He cautioned Rithy that a branch could fall in the night from a tree.

"But I found out later, it wasn't about the tree falling down. It's about a spirit living nearby," Rithy told me.

That night, it rained. Rithy thought his tent was waterproof, but driving raindrops found their way in. All his clothes were soaked. In the early hours of the morning, he felt like he was hallucinating. Outside, he heard someone shouting. His arms were covered in goose bumps. He unzipped the tent and stuck his head out into the rain. "I saw fire flying," Rithy told me. It looked like someone was twirling a bamboo stick of fire in the air, about a hundred meters away.

"I started to feel strange," Rithy said. But then, somehow, he pulled himself together. He told himself, *I'm okay. Nothing is going to hurt me.* A few years before that, Rithy had lived in a pagoda and practiced meditation. It taught him to be confident, to have willpower, no matter the situation. "That helped a lot."

In the morning, the monk walked up the hillside to ask Rithy how he had slept. In the light of day it was just a mountain. There were no other people. Nothing. No road on the other side where the fire had been dancing.

Rithy told the monk about what he had seen in the storm. The monk told him that ten years before, when he first moved to the mountains from Phnom Penh, he practiced walking meditation and chanting. The spirits tried to convince him to jump off the cliff with them. When he said no, they disappeared. But sometimes, they come back.

Rithy walked back down the hillside and ate lunch with the monks. He helped them build a small shed. The second night, he developed a sense of where not to sleep. He trusted his intuition. "Even the same rain, it doesn't feel the same," he told me. In the morning, he packed his tent and walked away from the trees. He told the spirits, "Thank you for giving me a good night."

After this encounter, Rithy started believing in spirits. They aren't ghosts, as he describes it, but "there's some spirit all over." In other words, "People don't just die and leave. Their body may be gone, but their energy—some sort of power—stays around for a while."

In the city, it's harder to feel connected to the spirit world. Rithy works as an entrepreneur and computer programmer at SmallWorld Cambodia, a collaborative workspace for young entrepreneurs. Technology, he believes, is "all just energy." And some of that power is formed in nature. When we met in 2016, Rithy tried to go to the mountains as much as possible, "not just to take a selfie" but to regain energy or release stress. The practice of observation and appreciation helps him recenter.

As we spoke, it poured. The street stayed flooded for an hour.

"Work in Phnom Penh is getting busier," he said. "I kind of miss my life back then."

MONKEYS

In 2016, Cambodia experienced its worst drought in decades.[2] The earth at the bottom of lakes cracked into fragments. Water resources were low. I met Panha Suon, a computer programmer, in Phnom Penh. "It was too hot, no water to drink," Panha told me.

Animals suffered, too. In Siem Reap, a female elephant known for ferrying tourists around Angkor Wat collapsed from heat stroke and died.[3] In Battambang, a man named Touch Tren found the bodies of thirty black monkeys in the Veal Don Om forest, all dead from dehydration. One of the monkeys, a mother, died with her child in her arms. This made Touch cry. He collected water from his village, already precious, and brought it to the forest for the surviving monkeys to drink.[4]

"For the last couple of years, it has been getting hotter and hotter. And we all can feel that," Panha said. "It's not going to be good. We have to do something," he added. "It's really up to us if we want to turn back, to walk away."

WASTE

Prach SoengChealy is a volunteer with Young Eco Ambassador Cambodia, a community of young Cambodians who want to take action on the environment. He teaches school kids about plastic waste and the flow of contaminants into the water supply.

"The majority of garbage in Phnom Penh drains into the river," he told me. "This is the same river that we take water for daily usage, which we call tap water."

Some sewage flows into the river, too. "The fish that live in the river eat that waste, especially plastic waste. That impacts the environment and the health of people who eat the fish," Prach added. "Many people who live near the dirty sewage get respiratory problems and diarrhea."

What we do to the bodies of water that surround us, ultimately, we do to ourselves.

RETURN

I tried to connect with other storytellers but quickly realized that I needed a break. I found out that I had an amoeba in my intestine and made the decision, after being on the road for almost two years, to come home to heal. I flew home to New Hampshire, where I took meds to vanquish the amoeba. I saw a therapist and learned what panic attacks were. Knowing what was happening in my body, and my cues for panic, gave me a sense of ownership over myself again.

I felt like a failure. My plan was to ride my bicycle around the world. But I soon realized I could travel to all the places that would be more difficult to visit on two wheels. The unifying metaphor was still there. Cycling, a way of moving. But also, sometimes, caring for yourself, preserving yourself, means learning when to let go.

And of course, my brush with violence didn't occur while I was on my bicycle. It happened when I was confined inside a van. If anything, I feel safer and freer on my bike than I do most anywhere else. Long van rides offer little opportunity for escape. On the bicycle, if someone bothers me, I ride away.

After I returned home I stopped, for a time, traveling by bicycle. I leaned into the new journeys. I let the stories lead. I came to understand my own breath. I started to learn who I was without the bike to define me.

For a time, I was afraid of being trapped inside any vehicle similar to the van that carted me across the border. Then, I wasn't anymore. I kept moving. Listening to other people, in some cases, involves closely listening first to ourselves. The man who put his hands on my throat wanted me to be silent, but I guess I won.

9

CHINA

SOUTH

From here, the journey became more jagged, less linear. After working as a rowing coach and taking time to recenter, I said yes to traveling whenever and wherever I could. In 2017, I was invited to fly to China as part of a delegation of former collegiate rowers who would compete in a regatta against local teams.

I arrived a week before the races in order to listen, and hired Eva, a recent college graduate who had studied English, to help me translate. Eva wrote a version of the "tell me a story" sign in Mandarin on a piece of cardboard that we scavenged from a stationery store (by this time, I had learned to always travel with a Sharpie). We met up in the mornings outside the apartment where I was couch-surfing and took public transportation to different parts of the city, pausing to speak with anyone who had a story to share.

Within the first few days, I learned that China's water resources are not equally distributed. One third of the country's nearly 1.4 billion people are concentrated in the north, which has less water than the south. This problem is not new. In 1952, Chairman Mao Zedong proposed moving water from the plentiful south to the more arid north. "The South has plenty of water and the North lacks it, so if possible, why not borrow some?" he asked.[1]

The effort to create long-distance water transfer projects has been more than half a century in the making. When completed, the South-North Water Transfer Project will link four of China's main rivers—the Yangtze, Yellow, Huaihe, and Haihe—and divert water along three canals: the East, Central, and West routes.[2]

I first heard about this large-scale movement from Frank, a teenager from Chengdu who was visiting Du Fu Thatched Cottage, a park and museum that honors the Tang Dynasty poet known for conveying the emotional impact of political and social issues of the time. Frank sat on a set of steps in Du Fu's lush gardens, explaining population density and demand in front of a pond brimming with lotus flowers. A butterfly flew past our noses.

"Sometimes our water resources aren't enough," Frank said. "I think people in the North should have the sense to save our water."

Both the East and Central canals have been completed. The western section is likely to be under construction until 2050. By that time, the Chinese government estimates that 44.8 billion cubic meters of water per year will travel from the south to the thirsty north.[3] The project's $62 billion price tag is more than twice the cost of the Three Gorges Dam, the world's largest hydroelectric power project, which began in 1994 and has flooded hundreds of square miles of land, creating a reservoir on the Yangtze river.

FRUIT

I met Ren Jiayou at the Wangjiang campus of Sichuan University in Chengdu. Ren, an undergraduate student, grew up in a farming community on the banks of the Jinsha River. As a child, Ren told me, "I could drink the water of the Jinsha River. I didn't worry about its quality."

But as he got older, the steel industry expanded. The river quality deteriorated. As fruit farming became more popular, farmers applied pesticides to protect their crops. Now, Ren wouldn't dare to drink the water in his own hometown.

WISHES

At the top of Mount Qingcheng in Dujiangyan, a city just northwest of Chengdu, incense burned thickly at the steps of a Taoist temple. Eva and I took a moment to catch our breath—we had hiked several hours to reach this summit. Seeing my cardboard sign, a woman named Weimin approached us. She said that she comes from the coastal city of Jinzhou, in Liaoning Province. In Jinzhou, there is an island just offshore called Bijia Mountain. During most of the day, when the tide is high enough, Bijia Mountain is inaccessible except by boat. When the tide is low, though, a cobblestone pathway appears, just wide enough to walk across. There is a Taoist shrine on the island in honor of Pangu, a mythological figure who separated the sky and the earth, chiseled out valleys, and stacked up mountains.

As a child, Weimin could walk to Bijia Mountain at low tide. As she grew older, Jinzhou Bay was built up, and the construction changed the water flow. The pathway to the island didn't show up anymore. If people wanted to reach Bijia Mountain, they had to take a boat. "Later on, the concept of environmental protection became popular," she told me. "Due to the island being a world wonder, a lot of resources were invested to remove the sand and bring the footpath back to the surface." Soon the pathway returned, complete with its sand, stones, moss, algae, shellfish, and crabs. There's a saying, Weimin said, that if you're a student, you should make a wish for good academic performance on top of the mountain.

One eighty-nine-year-old woman in the community walked to the island every day to voluntarily light candles for other people for luck. "After making a wish, you write down your exam date, and the old lady will light a candle for you on that specific day," Weimin said. The woman had been doing this since she was younger, and it became a habit of hers to make a return journey to the island each day, following the tides. When the footpath was engulfed by the sea, her children took her to the island on a fishing boat. "But now she is happy that the footpath resurfaced," Weimin said.

This is the role that elders have—to carry the flame for everyone else.

COTTON

Ahmatjan Rouzi is a Uighur from Ürümqi, a city in the northwestern region of Xinjiang Province. As a research associate at Xinjiang University, he focuses on water management in western China. We met in Denmark at a conference focused on the UN Sustainable Development Goals.

The Tarim River, Ahmatjan told me, is one of the largest inland rivers in China, but due to climate change and population growth, riparian forests are disappearing and droughts are becoming more common. At the extreme end of this spectrum is desertification, sandstorms, and salinization. And sandstorms don't just stay in the desert—they threaten the city as well.

"One of the main reasons for this problem is that we have cotton as a cash crop," Ahmatjan said. In his research, he works on how to switch to other crops that are better adapted to local conditions. "My hope is to provide policy recommendations from our research to the local government," he said. "I hope the regional government will have a more holistic approach to diversify the region's agriculture and, at the same time, invest more money on environmental protection and improve water efficiency."

MEMORY

I met Ren Hu, a PhD student at the University of Wollongong, when I was cycling through Australia, but the story he wanted to share was about his hometown, close to the city of Nanjing. "When I was a kid, about seven or eight years old, in my hometown, the snow in winter could be very thick," he told me, "and everyone made a very big snowman." The memory of those winters made him smile.

But a few months earlier, he had returned to his hometown in the winter. Now, on the rare occasions when snowflakes fall, they melt without accumulating. Ren now thinks of snow "like an endangered animal, because it's very rare in my hometown," he said. "My memories became a fairy tale."

GREAT WALL

On a high-speed train from Chengdu to Beijing, I pressed my nose to the window. The colors blurred: gray concrete to lush greens to lanky beams of steel.

Partway through the fourteen-hour trip, the people next to me made noodles and tall thermoses of tea. A screen in the passenger car clocked our speed at 304 kilometers per hour.

In Beijing, I met Jo Darrington, a tall Singaporean with a big smile and a quick laugh. Her black hair was dyed with streaks of blond and green. We traveled together to the Great Wall of China and climbed a hill to a wild stretch of the wall called Xiangshuihu, which Jo said means "sound" or "echo of water." I followed her up crumbling stone steps where plants grew through the cracks. The hills below us were jagged and green. It was the kind of place where you could feel the weight of the past. Clouds above us threatened rain.

We stopped at a point overlooking a dramatic valley, and Jo told me a folktale. The story began six hundred years ago, when soldiers were building this stretch of the Great Wall. It was a dry season, and the laboring soldiers were suffering from thirst.

"One day, a woman appeared and she gave them some water. And then she mysteriously disappeared," Jo said. Legend has it that the woman was Bodhisattva Guanyin, a figure in Buddhist lore who carries a jar of pure water in her left hand and a willow branch in her right.

But the generals at the wall didn't recognize Bodhisattva Guanyin. They thought she was a local villager and took the water gratefully. The generals didn't drink it themselves. "They passed it on to the soldiers, and the soldiers said, 'Oh, you know, you need it more than us,' and passed the water on to the villagers," Jo told me. The villagers gave the water to the women, who passed it on to the children.

Even in a drought, a time when scarcity can be a source of escalating tension, everyone agreed that women and children are the most vulnerable. This generosity pleased Bodhisattva Guanyin. "And so, on seeing this, the mysterious woman stamped her foot and created a lake," Jo said. "And then there was plenty of water for everyone."

DAMS

When we spoke in 2017 in Beijing, Stephanie Jensen-Cormier was the China program director for International Rivers, an environmental organization that started in 1985 and now works to address the legacy of dams in more than sixty countries. We sat side by side on a bench in Chaoyang Park.

"In China, thirty years ago, there were fifty thousand rivers. This is a number from the Chinese government, from the Bureau of Statistics. And the same Bureau of Statistics now says that there are twenty-three thousand rivers left in China, so that means that more than half of the rivers have completely disappeared," she explained. "This is due to pollution, contamination, and also because the rivers have become dry. So rivers have been overexploited by industry, by agriculture."

In order to mitigate the impacts of climate change, China has been building hydropower dams. International Rivers has focused their work on trying to protect some of the last free-flowing rivers in China. One of those is the Nu River in southwestern China. It starts in the Tibetan plateau, flows through Yunnan Province, and then into Myanmar and Thailand, where it's known as the Salween River. In addition to being an important river regionally, the region has some of the highest biodiversity in China, making it a crucial area to protect.

China's centralized system plans in both long- and short-term intervals. Every five years, the government announces a list of priorities. From 2010 to 2015, one of those plans was to build a cascade of dams along the Nu River. During that time, International Rivers partnered with Chinese environmental NGOs working in Yunnan Province to advocate against their construction.

"We did a lot of research on the geology of the region and why it didn't make sense geologically to build new dams," Stephanie said. They also brought in botanists to research the biodiversity of the area. When the next five-year plan was announced, there was no mention of building dams on the Nu River. "That is, I think, a big success," she added. While the change was likely not entirely due to their advocacy, it did play a small part.

This is especially important in China, where protest is generally not a viable strategy. "Civil society is still not built up and not empowered to be making big statements like that. The government wants to maintain control and wants people to be happy but is very quick at squashing any kind of rebellion or protest," Stephanie said. Instead, the best practice is to work with the government, provide them with research briefs, and never criticize them publicly.

China's biggest hydropower companies are state-owned enterprises. The government's "Going Out" policy in the late 1990s encouraged the companies to build projects outside China, both to turn a profit and to build better ties with governments abroad. As part of that, Chinese companies were involved in building hundreds of large dams internationally.

Hydropower, despite its green reputation, is not climate-neutral. In the creation of a dam, a large area is flooded to create a reservoir. These areas usually have a large amount of biomass and vegetation. When that area is flooded, the vegetation disintegrates into the water and releases greenhouse gases. It can be hard to visualize the emissions, but they are there.

Another concern is the ecological impact. "We should imagine that rivers are the arteries of the planet. Rivers are so important to the healthy functioning of the entire system that we're living in that stopping the flow of a river has detrimental impacts further down," Stephanie emphasized. Among other things, creating a dam means that fish and other species who migrate along that corridor cannot pass.

The social impacts of a dam can also be detrimental, most acutely for people whose livelihoods depend on the river. Millions of people along the Mekong River are subsistence farmers and fishermen. Their protein comes from the fish that they catch, and their produce comes from river gardens on the banks of the Mekong. "If you alter the flow of the river or stop the flow of the river, you're having really big impacts on all of these people's livelihoods and survival," Stephanie said. Many of these communities exist specifically because the soil is fertile, or because their location gives them access to transportation by boat. For generations, communities buried their ancestors there. "To them, it's really important to stay where your ancestors were, and it's also unthinkable to leave your ancestors behind. So any resettlement issues are very complicated," Stephanie said. The ancestors cannot be moved, and there is no compensation that the living want to accept for the dead. "A tree, like a fruit tree or something that will get flooded, there are ways to try and figure out compensation for that. But some of the more intangible human, cultural heritage is really difficult to think about," she added.

One of International River's strategies was to create a benchmarking report where they ranked the social, environmental, and risk management policies of seven large Chinese hydropower companies who work in other parts of the world. They gave each company a score against twenty-six different indicators, and then ranked the companies. "The companies that did really poorly got upset about it, and the company that did the best asked for a recommendation letter from us," Stephanie said. Even those that did poorly were eager to learn how they could improve their projects. All the dams were in the final stages of completion, but there were still small steps that could be taken to ensure better environmental and social outcomes.

10

KAZAKHSTAN

DOORWAYS

In 2018, a classmate from college, Didar Kul-Mukhammed, invited me to be a part of Go Viral, a festival in Almaty, Kazakhstan, sponsored by the US Embassy and Consulate and focused on innovation and ideas of all stripes. In exchange for being filmed for a short documentary, the festival organizers furnished me with a translator, Qanat, who accompanied me on a ten-day journey to Lake Balkhash and across the border into Bishkek, Kyrgyzstan, to record stories about water and climate change in local languages. The videographer, Sardar, followed every movement with his lens, and a driver transported us from place to place. This was my first time documenting stories while accompanied by three men—a far cry from traveling alone on my bicycle and letting happenstance dictate my route. Still, the essence of the journey was the same: I was out there to listen, I just had a bit more help than I was used to.

Simultaneous translation was sometimes frustrating, sometimes amazing, and always a learning experience. I would ask a question, Qanat would translate, and then we went back and forth. With practice, we became proficient in the art of linguistic triangulation. Step one: listen. Step two: listen again.

Cardboard, blissfully, is everywhere. When we arrived in Balkhash, a city on the northern shore of a lake that is half salty, half fresh, a fruit seller on a sidewalk lent me a box and a pair of scissors. Qanat wrote down the words in Russian, and I filled each letter in on the new piece of cardboard. The storytelling sign made its Russian debut. Soviet infrastructure, rectangular and orderly, dominated the urban landscape. People didn't trust the utilities and bought bottled water to drink.

Each doorway I walked through was a relic of central planning. The economy of Balkhash centers around the copper smelter. Its smokestacks figured prominently on the horizon. When the wind was blowing toward the city, it made me cough. My eyes watered. One person I met on the banks of the lake told me that you get used to the air. After a few weeks, he didn't smell it at all.

RIVERS

Daulet Kozhakov is a historian and curator at Balkhash City Museum. Qanat and I followed him as we walked past a diorama of a marshy ecosystem, complete with birds, fish, and tall grasses. "We've been destroying this," Daulet told me through Qanat, gesturing at the plants and animals. "Certain birds are leaving and fishes are dying out. The freshwater part of Balkhash is becoming saltier." Rivers that were dammed in the 1960s and '70s no longer supply water to the lake.

The culprit: agriculture. Soviet canals are inefficient. In Israel, Daulet said, a cucumber requires some forty times less water to grow than it does in Kazakhstan. Drip irrigation simply isn't practiced here; there has been no innovation in water management for centuries.

Daulet likened this phase of history to just after Europe introduced industrialization. "They had wild capitalism. They completely destroyed the Rhine River," he said. "But when their nations understood better, and when they grew up as a society, they took care of the Rhine. Right now the ecology is much better." In Kazakhstan, Daulet sees this same kind of post-Soviet "wild capitalism" taking root. "Everyone's concern is only money, how to get rich without, of course, caring for nature." His concern is for the ecosystems that will be threatened in the time it takes for people to catch up. "The only solution is that we have to slow down our appetite for money," he concluded.

SNOW

One evening, while wandering on the lakeshore, we met Nurkhan Ashikbaev, a man with crinkles around his eyes and a wide stance that projected confidence. Nurkhan has worked at the copper factory in Balkhash since 1995, when he moved from the Karaganda region, a grassy steppe, in search of work. "In small villages in the Soviet Union, you couldn't do anything much," he told me. "I had to move to a bigger place where I could find work." Now, his days start at eight a.m. In the factory, he operates heavy machinery to boil the copper in a furnace. "The molten copper flows like water," Nurkhan said. Another machine rotates 220 kilograms of copper at a time, cooling the metal before it's sent to the electrochemical department. They use water from the lake to cool down the copper. "It's just a pure, raw, mechanical, physical job," he said. "Work is work."

At the end of the workday, Nurkhan likes to take a stroll by the water with his family to relax. He worries about the fish—some fishermen are careless and discard their nets in the water. "The amount of fish that people catch with nets is not regulated," he lamented. "When I go fishing, the fish are smaller." Without regulation, he is concerned that overfishing will deplete the stock.

In the winter, the lake freezes completely. People go walking and skating on the surface. Even cars can drive over it. Ice fishermen drill holes to find their catch. And although the ice is consistent, there is less snow than he remembers from the 1980s and '90s. But when it does snow, play becomes a priority. Nurkhan makes snowmen with his seven-year-old daughter. They throw snowballs. "We are happy when it's snowing so we can play around and have fun," he said.

The lake also figures prominently in their summer recreation. His older sons, ages twenty-three and nineteen, all learned to swim in the lake. Nurkhan was a gentle teacher. He didn't throw his kids in. Instead, he held under their backs as they looked up, showing them how to make certain movements as they relaxed. "Then I let go just a little bit," he said, "and they swam."

FUTURE

I met Alisa on the road leading to Lake Balkhash, where she was drying fish in the sun. "Our situation is getting worse and worse every year," she said. "The water is salty. We cannot drink it so we buy bottled water. It's expensive." She has also noticed that fewer and fewer fish are caught each year. "The government is not really caring about us," she said. "This is the only way I can earn, a little entrepreneurship. I'm selling fish."

In the future, Alisa hopes for a park, a place to relax. Even though there are limited opportunities for business, "I still don't want to leave this place. I like it. I was born and raised here and it is my place," she said. "I don't want to move anywhere. I want to stay here and I want this place to get better."

AIR

Back in Almaty, I connected with Asya Tulesova, an environmental activist. We sat on a bench in a public park, breathing the multilayered moisture of an impending storm. Asya told me that the mountains that skirt Almaty, while beautiful, trap pollution within the city. "Whatever we produce stays here, and doesn't leave," she said. Air pollution in Almaty has been a concern since the 1970s, and it's become more pronounced in recent years with population growth, more cars, and a central heating system that uses coal to heat homes.

In order to draw attention to this problem, Asya started Almaty Urban Air, an air quality monitoring project. A network of sensors informs people how bad the air quality is in the city on any given day. Asya cocreated an app that anyone can download and check on an hourly basis.

"The problem for our cities is that data is not easily accessible," she explained. "We didn't expect people to be so obsessed with it. It had a viral effect. People didn't expect the numbers to be so high." They released the app in December 2015, when the heating season was at its peak. Aside from drawing attention to the problem, her goal is to work with NGOs in the Almaty area to come up with actionable recommendations for the city government.

"It's so easy today to do a project like this, because technology is available. It's just a matter of being bold and a bit courageous," she said.

SNOW LEOPARDS AND BIKE LANES

Kok Zhailau is an alpine valley that's a quick bus ride away from Almaty and part of Ile-Alatau National Park. The area harbors biodiversity and supplies fresh air and clean water to the city. It's peaceful, too. Almatians can hike trails within the park that lead far away from any road or construction site. Svetlana Spatar, a leader of the organizations Green Salvation and VeloAlmaty, takes respite from the city in these highlands. While there, "You have a deep feeling that you are in true nature," Svetlana told me. "It's very accessible by many people, young and old, rich and poor. Any layer of society could go there." This location is one of a few natural areas where snow leopards, a symbol of Almaty, live.

Within the last fifteen years, the city government announced a plan to develop Kok Zhailau to bring infrastructure for tourism, skiing, and skating. Svetlana helped lead opposition to the plan, certain that it would disturb the unique and fragile ecological system.

"Another thing that outraged us is that a huge amount of money was supposed to be dedicated from our local budget just for that construction. With so many troubles and issues within the city, we thought that this money would have been used much better if we just did something inside the city," she said. She helped gather more than seventeen thousand signatures of people against building infrastructure in Kok Zhailau, and she held workshops and flash mobs in the city to help educate people about the problem. She also wrote letters to the UN, appealing for protection from the international conventions on biodiversity. Finally, the construction project stopped when the money dried up. Their vocal activism had helped. A few years later, though, the local government revived the plan and found an outside investor. "We are still fighting. We didn't give up," Svetlana told me. The valley, she said, "is a sacred thing." That first campaign solidified the community and made them more active and outspoken.

In the early 2000s, Svetlana's husband gave her a bicycle as a gift. She had recently returned from a trip in the Netherlands, where she saw how bicycle infrastructure could transform a city. Svetlana wanted something similar at home. At the time, there was a small group of cyclists in Almaty who would gather to ride together and talk about infrastructure issues. They started a movement called VeloAlmaty.

When the group started, cycling was somewhat taboo. The predominant attitude was "if you're riding a bicycle, that means you're poor and don't have enough money to buy a car," Svetlana said. "Any person that has a certain self-respect would not ride a bicycle. And there was no infrastructure supporting it at all."

The group's first action, in 2007, was to write a letter to the mayor asking him to consider constructing bicycle lanes. They compiled almost a thousand signatures. The mayor replied but was noncommittal. "We stayed optimistic. We didn't give up," Svetlana said.

Next, they went to a higher body of government. They contacted committees and ministries in Nur-Sultan, the nation's capital city, with a connection to transportation or ecology. To Svetlana's surprise, it worked. The Ministry of Transport and Communications took interest in their request. The vice minister wrote a letter to the mayor asking that he work with VeloAlmaty.

The result was minuscule. The city built two kilometers of a bike lane. "But it was a kind of revolution. It was a turning point," Svetlana said. The infrastructure grew from there. Now there are more than sixty kilometers of bike lanes in the city, and riding a bike is less taboo. VeloAlmaty is working with local businesses to create bicycle parking. They also host workshops to popularize riding bicycles for children and elderly people and meet with drivers to advocate for respecting bicyclists.

Within just a decade, "We've changed our community," she said. "You can see many people bicycling, including people wearing suits, CEOs, and owners of businesses." And every bike is helping to improve the quality of Almaty's air, for everyone.

DEGRADATION

Toward the end of the Go Viral Festival, I met a doctor in Almaty who didn't want to share his name. Years ago, he worked on a humanitarian aid project in the Aral Sea region. "To really understand the story of the Aral Sea, you've got to go back to the time when Stalin was in charge of the Soviet Union," he told me. Stalin felt that the Soviet Union shouldn't be importing cotton and should instead have a domestic cotton industry to supply its own needs. "He, for reasons that make no sense to anyone who knows anything about agriculture, decided that Uzbekistan would be the place to do it."

The climate and conditions in Uzbekistan were, in the doctor's words, "completely wrong for cotton. It's a very arid climate." Cotton needs a lot of water to survive. In order to support all the cotton fields that Stalin wanted in Uzbekistan, the USSR created an elaborate irrigation system. The nearest source of water was the Aral Sea.

"Now, the Soviet Union being as it was, no one was better at anything than anyone else. So the guy who was put in charge of this project, in fact, had no real knowledge or background in the appropriate engineering," the doctor told me. "I've actually spoken with him. He passed away, but at the time, he felt very remorseful about his role in it because he was forced into this horrible position of doing something that he didn't know how to do properly. He knew he was doing it wrong. But he also knew that he didn't have a choice, and the consequences for him and his family would be very dire if he didn't cooperate."

As a result, the canals weren't lined or managed properly. Most of the water was lost before even making it to the fields. And huge amounts of chemical fertilizers and pesticides were applied. "Over time, what you're creating essentially is a toxic salt marsh," the doctor said. The water table shifted. The sea level began to go down. Eventually, the irrigation channels dried up, too.

By the time the doctor was treating the local population, he saw high levels of unusual cancers and autoimmune disorders. Agriculture had been destroyed, and the waterways were no longer viable. A fishing village that was once a popular vacation site in the Soviet Union is now a ghost town. The robust fishing industry and cannery have disappeared. "It's just desolate, with really nothing to support the economy," the doctor said. "You can stand on what used to be the edge of the sea and look out into what used to be a seabed. It's tragic."

The doctor sounded awestruck. "I couldn't imagine that human beings were capable of destruction on that scale until I actually saw it. And it just blew me away. It was terrifying," he told me. "For anyone who wants to deny the consequences of human action, take a look. Humans are fully capable of destroying huge sections of the planet and making it unlivable."

BROTHER

Some stories shoot through you like arrows. I got to know Ashan Illah in the final hours of Go Viral, as organizers cleaned up the room and folded up chairs. All weekend, I'd worn my cardboard sign, asking for stories. Ashan approached me. He was twentysomething, clad in glasses, jeans, and a button-up shirt. I could tell by the pattern of his breath, the way his voice broke as he spoke, that he was nervous. He said he wanted to tell his story for the first time.

"Where I was born, we did not have enough water," he began in halting English. He was born in a rural home in Afghanistan, alongside a river. In order to make the river water safe to drink, the family kept a fire in the middle of the room and boiled the water there.

Ashan told me that one day, when he was six or seven years old—his brother was a toddler—his parents were away, and the boys were alone with the boiling water. As he collected himself to tell me the next part of the story, Ashan was visibly shaking.

"My . . . my little, little brother came and . . . just . . . pulled it," he said, faltering. "And it dropped on him." The boiling bucket of water toppled onto the toddler. Ashan's brother died that day.

The experience haunts him. When he drinks a glass of water, he thinks about how, if his family had only had clean drinking water, his brother would still be alive. Years later, Ashan is still trying to process this tragedy. He dreams of developing the green technology that will bring safe drinking water to Central Asia.

Later, his family moved to Kazakhstan, and he went to school to study civil engineering. "If I had to lose my brother," he said, "maybe someone else won't be lost in the future, because the water was clearer." For now, though, access to water in his hometown is still much the same.

11

KYRGYZSTAN

ISSYK KUL

My time in Kyrgyzstan was brief but beautiful. Qanat, Sardar, and I ventured across the border from Almaty to Bishkek. In between wandering around the city with my cardboard sign and recording audio stories, I gave two presentations at Chicken Star, a hybrid fried chicken, coffee, and art establishment known for hosting events that brought a cross-section of locals and foreigners together. Chicken Star's founder, Chihoon Jeong, is the kind of person who can intuit what kind of drink you need before you even know that you need it. For me, it was a hot elixir of ginger, lemon, and honey. What a gift.

Each culture has a unique tradition of storytelling, but there are commonalities. Origin stories about how different bodies of water were created are common. I heard one of these water myths from Nurperi, a ten-year-old girl in the fourth grade. We met while she was at one of Bishkek's public parks with her older siblings. "There is a legend about Lake Issyk Kul," she began. Two boys fell in love with a girl. They fought over her, and then they killed each other. The girl was so sad that she cried an entire lake. "That's why it's so salty," she told me. The lake is known for its ability to heal people. "It's really big and really beautiful and shiny," she said.

Nurperi doesn't remember who told her the story. And that's the beautiful thing about myths—there are as many ways to tell them as there are people on the planet. Water is not just a natural resource. It seeps through our culture and lives in the stories we tell each other.

ALAKIYIZ

Asselle Orozbakova works as a guide at the National Museum of Fine Arts in Bishkek, an imposing, square-shaped building with tall ceilings and a faintly musty scent. In a room of Kyrgyz embroidery and felt rugs, she showed me a piece called *Contemplation*, made by Jumabai Umetov in 1980. The piece, made of felted wool, is called an *alakiyiz*. Originally, these pieces of felted wool were used on the floors of yurts, to keep them warm and protect from the wind.

"There would be no *alakiyiz* without water," Aselle told me. The technique for creating one is elaborate, involving rolling a mixture of wool and immersing it in hot water boiled from a mountain stream. "What is interesting in the making of *alakiyiz* is that the entire family would be involved because you need some power and lots of people to produce such a piece of art." The resulting piece of wool is warm and flexible. Jumabai Umetov was among the first artists to revive this practice, and to propose hanging the *alakiyiz* from a wall.

The resulting textile, bold and colorful, carries the whisper of the river into the silent gallery.

ENGINEER

In Bishkek I also met Aidai Turdakunova, a sixteen-year-old student, who had come to Chicken Star. She told me that her goal is to become an environmental engineer and address problems related to both water and climate change. Aidai's plan is to study abroad and work for the United Nations "to protect the world and to protect my country as well," she told me. "There are many kinds of problems in the world, and I want to change this for the better."

Her dad also wanted to become an environmental engineer when he was younger, but he didn't pursue that dream. "He had some problems with chemistry," Aidai told me. The easiest solution, as she sees it, has only two wheels. "It will be better if we use bicycles," she said.

One of her top concerns is pollution from transportation—especially cars.

DIVISION

I met Omurbek, a sixty-two-year-old man in a tracksuit, while he was taking his afternoon stroll in a leafy park near Bishkek's center. In his view, "Water is just one body," inseparable from the other bodies of water that it flows into. Borders are superfluous. "We should not say water in Kyrgyzstan and in Kazakhstan and other countries. It's all connected as one body of water," he said.

Anything we do to the water we do to ourselves. "As we change the condition of water, it changes not only just the water itself, but it actually changes us," Omurbek added, putting his hand over his brow to shield his eyes from the sun. "We should treat everything around us as a live being." This includes water, which he views as alive.

In 2014, Kyrgyzstan suffered its worst drought in decades.[1] The rivers supplying Bishkek have since been strained by population demand. "As the population grew, we needed more water. And, of course, agriculture needs water, too." Completing this thought, he ambled away, leaving me to contemplate the network of droplets that keeps the city alive.

EPIC OF MANAS

One evening, over plates of cheese sticks and fried rice, Asan Zhunusbekov told me about the *Epic of Manas*, which, at over five hundred thousand lines, is among the longest stories in the world.[2] (*The Iliad*, for comparison, is under sixteen thousand lines). The Kyrgyz epic was passed through generations orally, recounting the story of a man who united the Kyrgyz nation. The one thousandth anniversary of the epic was celebrated in 1995.

"He's a son of Sky and Earth. He's a true spirit," Asan told me. As a child, Manas was reckless, but "he had special abilities, like amazing, tremendous leadership skills." The story includes mystical elements; Manas had a flying horse. His wife could turn into a swan and fly away. Each retelling of the epic involves verbal improvisation. "It's poetry," Asan said.

Before the formation of the Soviet Union, Asan told me, Kyrgyz people didn't swim in the lakes. They were seen as sacred—something to be revered, but not entered. "Traditional Kyrgyz culture respects nature more than we do now in Kyrgyzstan." He recounted two ancient sayings: first, "There is no sin in moving water." In other words, water purifies itself. And second, "Don't pour water where it will not be absorbed." In other words, attention is precious. Don't waste it on things that will yield nothing.

Although Kyrgyzstan has abundant water compared to other countries in the region, its old irrigation systems are inefficient, and much of the water, Asan told me, is lost before it reaches its destination. "I think we need to raise awareness that it's so important to save every drop of water," he said.

Older people, Asan added, have a lot to say about climate change—the weather in Kyrgyzstan has become "more polarized. Winters became colder and summers became hotter." This tracks with his own memory of the seasons. "I think that is one of the signs telling us there is something wrong happening in the world."

But he recognizes how difficult it is to quantify something as vast as climate change within the scope of a single human life. "It happens so slowly. It cannot fit in a box," he said. "It's a spectrum. It happens continuously, and those kinds of things we need to measure from generation to generation." Some signs, like glaciers melting, are obvious. "This is how water is stored in Kyrgyzstan, in the mountains, glaciers, and from year to year they're shrinking and shrinking in size," he added. "We have to think local and we have to act local, because the situation is so diverse everywhere."

12

MOROCCO: UN CLIMATE TALKS

SOIL

Before stepping into COP22, the 2016 UN Climate Change Conference held in Marrakesh, Morocco, we found a patch of grass. Surrounded by the flagpoles representing the 196 countries present in the negotiations, someone unfurled a blue flag, the marble of the Earth in the center, and placed it on the ground.[1] It was raining lightly. Dark-red clay caked our shoes. One of the thirteen in our youth delegation group started to sing. Another prayed. We closed our eyes.

Then, outside one of the tents, we joined a line that snaked past an airport-style security check. We emptied our pockets, took off our shoes, and posed in front of a blue wall for a badge photo printed with the logo of the United Nations Framework Convention on Climate Change. I entered the space as an observer—no negotiating power, just the power to listen.

As we walked, dried clumps of soil fell from the soles of our shoes onto the carpet.

"Look," Kailea Frederick, a member of the delegation, said, pointing. "We're bringing the Earth into the UN."

NEGOTIATIONS

Since 1995, the Conference of the Parties, the decision-making body of the United Nations Framework Convention on Climate Change, has been gathering annually to discuss how best to respond, internationally, to climate change. The first COP was held in Berlin.

COP22's two weeks of negotiations, November 7 through 19, 2016, brought together more than 22,500 participants, including nearly 15,800 government officials, 5,400 representatives of UN bodies and agencies, civil society organizations, and 1,200 members of the media.[2] During this time, the parties adopted thirty-five resolutions related to long-term finance, capacity building, and addressing loss and damage associated with climate change. There was hope, albeit in the form of bureaucracy.

When I was in Tuvalu in 2014 and spoke with Soseala Tinilau, he had just returned from COP20 in Lima, Peru. "It seems like all the parties really want to have a legally binding agreement come COP21 in Paris next year," he told me. "Our hope as a small island state and vulnerable country is to have a legally binding agreement to continue the Kyoto Protocol as a mandate for trying to reduce emissions so that we can help target 1.5 degrees Celsius."

COP20 in Lima, Soseala told me, "laid a foundation in order to come up with a pretty sound agreement come COP21 in Paris next year." He was optimistic that change would come.

ACTION

At COP21 in Paris, the Paris Agreement was adopted. The document aims to limit global warming to well below 2 degrees (Celsius) and peak global greenhouse gas emissions as soon as possible. Every five years, countries revisit their national pledges to track their progress.

More than 185 nations signed the Paris Agreement; the document entered into force on November 4, 2016. COP22 followed on the coattails of the Kigali Amendment, signed October 15, 2016, an agreement to phase out hydrofluorocarbons—a supercharged greenhouse gas used in air conditioners and refrigerators with a thousand times the heat-trapping potency of carbon dioxide.

The conference in Marrakesh, Morocco, aimed to plan concrete action toward achieving the goals of the Paris Agreement. These plans can be divided roughly into the following categories: adaptation, transparency, mitigation, loss and damages, technology transfer, and capacity building. I listened on the fringes: under tents, in line to enter press conferences, in the back corners of civil society rooms.

PREPARATION

I attended the 2016 UN climate talks in Morocco as a youth delegate with SustainUS, witnessing the complexity of international climate negotiations firsthand. SustainUS is a youth-led 501(c)(3) nonprofit organization that aims to advance justice and sustainability by empowering young people to engage in advocacy at the domestic and international levels. (Young people, by the organization's definition, are those under thirty).

In 2016, SustainUS brought thirteen youth climate justice activists from all corners of the country—from Hawaii to Utah to Navajo Nation to Virginia—to COP22. Morgan Curtis was our leader. Orion Camero, Becky Chung, Kayla DeVault, Remy Franklin, Kailea Frederick, Niria Alicia Garcia Torres, Ben Goloff, Noah Goodwin, Daniel Jubelirer, Brook Larsen, Dineen O'Rourke, and I were selected to come.

Two months before the negotiations—four months after I returned to the United States from Cambodia—our delegation came together for a training retreat at Canticle Farm, an intentional community in Oakland, California. Canticle is modeled after the ecophilosophy of Joanna Macy, an author and teacher whose work focuses on Buddhism, systems thinking, and deep ecology. She is best known for the Work That Reconnects, a workshop that helps people respond creatively to global crises as an alternative to feeling overwhelmed.

In this workshop, a facilitator leads a group of people through a spiral. There are four stops: coming from gratitude, honoring our pain for the world, seeing with new/ancient eyes, and going forth. Each section involved making eye contact, milling about a room, and imagining a continuum of ancestors—those who have come before, and those who will follow—flowing from each person present. The workshop model arose in the late 1970s, when nuclear warfare was a pressing concern.

Next, we learned about nonviolent communication, a way of understanding the needs of ourselves and others. The technique centers on deep listening and compassion. In order to put this into practice, the thirteen of us sat in a circle on the carpet. One by one, we spoke, saying whatever was on our minds: fears, feelings, all of it—the mess of a mind out loud. Scraps of paper at the center held the words of our collective needs: compassion, spontaneity, presence, shelter, validation, respect.[3] When a person finished speaking, we reached into the center of the circle to pick up a need, and then gave it to them. At times, I found it difficult to stay still. This was a different kind of listening than even I had done before. I had to quell my desire to respond.

Finally, we spent time with an easel pad, markers, and multicolored Post-it notes, creating the vision of what we wanted to accomplish together. Three goals emerged: active solidarity, storytelling, and community.

PEBBLES

At the conference, we spent our time under various white tents. Outside, the air was dry and red. Inside, it felt surreal. The site of COP22, the Bab Ighli village, was created in less than six months and meant to be deconstructed in its entirety after the conference was over. The fifty-five tents, pebbly gardens, carpets, pop-up photography exhibits, and grab-and-go restaurants all had an air of ephemerality. There was always a long line at the coffee stand. On the first few days it was cloudy. Then, the sky cleared and we could see the Atlas Mountains, distant on the horizon.

Some days, when there wasn't a pressing event, I let myself get lost in Marrakesh's maze of souks. The light filtered in from above, striated. Every city has its own scents. Marrakesh is new leather and fruit, punctuated by pollution from passing motorbikes.

NEWSROOM

It was at COP22 that I found my calling as a journalist. I was drawn to the press room, the rhythm of releases and briefings and swift deadlines. A hum.

I wasn't supposed to be there (my badge technically only allowed access to civil society spaces), but everyone was too busy to notice. I walked past booths from Agence France-Presse, Reuters, the Associated Press, *Le Matin*, *The Guardian*, *Democracy Now!*, and *Pacific Standard*. I didn't want to be anywhere else.

Standing shoulder to shoulder with these journalists in the back of press events, watching them conduct interviews and then type furiously, I realized that I wanted to join their ranks. At COP, I gathered snippets of stories about water and climate change, just as I had been doing for the two years prior, but I also started to pitch them to news outlets. I wanted in.

ELECTION

On November 8, 2016, as ballots were being counted back home in the United States, three members of the SustainUS delegation stayed up all night, making a "presidential to-do list." Historically, the United States has played a large role in the climate negotiations, and justly so; we are one of the largest polluters in the world. At the time, it seemed certain that Hillary Clinton would be elected. The idea was to hold her accountable to the climate movement around the world.

Ben, Dineen, and Orion painted with a toothbrush on a sheet. At some point, they gave up on the toothbrush and moved to bare fingers. Eleven items spilled onto the floor:

- Work for climate justice
- Respect Indigenous sovereignty
- Honor the treaties
- Cut corporate ties
- Zero by 2050
- Break free from fossil fuels
- Commit to a just transition

- No corporate trade deals
- Stop militarism
- End inhumane policies
- Protect water

Late at night—deep into making this banner—they started to get delirious. Someone had their laptop open to the *New York Times* needle, counting electoral college votes. Clinton's chances started to slim.

"We were getting more and more despondent—and determined," Ben recalled. "We had something concrete to do, this banner, even if the strategy didn't make sense." He continued to paint.

At dawn, the sound of the call to prayer bounced from nearby minarets. The banner was complete. And Trump was the president-elect. Daniel and Brooke went from room to room to wake the rest of the delegation. The moment "felt like a punch in the gut: heaviness, dread, despair," Daniel recalled after the fact. No one had imagined this outcome.

Remy had the idea to cross out the word "presidential" and make it the "people's to-do list." As the paint dried on the sheet, we gathered in a circle.

"Actually, this was always up to the people," Kailea reflected. "It's always going to be on us."

~

"Are you the Americans? Do you need a hug?"

Beatriz Azevêdo de Araújo, a youth delegate from Brazil, had attended every COP since COP19 in Warsaw, Poland, in 2013. She was warm and confident. Some youth delegates referred to her as their "afternoon espresso"; when things seemed hopeless, Beatriz and other members of the Brazilian delegation brought enthusiasm to an otherwise deadened space.

And on November 9, 2016, true to her reputation, Beatriz welcomed me with a hug and a small point of insight: governments are less powerful than the people they represent. "My first COP, I attended negotiations rooms," Beatriz told me, but in recent years, she hasn't set foot inside that space, a combination of choice and the fact that civil society members these days are allowed only limited access to the back rooms.

"From my perspective, it's kind of pointless," she said of the negotiations. "It's very easy to promise things and then go back home and do nothing." For Beatriz, in the absence of credible state action, any hope of meaningful action on climate change will require grassroots work by the public. "Youth here have to regain power and just take action," she said, "because I don't trust my government to do anything."

~

The United Nations, at its core, can feel cold. Removed. The vast majority of the attendees of COP are in suits. We were not them. We had something else to say.

The SustainUS delegation went outside, gathering near the display of flags from countries around the world. We stood in a circle alongside youth delegates from Switzerland, Canada, Mexico, New Zealand, Brazil, Vietnam, and elsewhere. Noah described it, later, as a "kinetic moment." We unfurled the people's to-do list. We were not alone.

"Donald Trump, we have a message for you," Orion Camero began. "We do not believe your hate needs to represent our country. We created this presidential to-do list, but it's up to people to decide what our country is about!"

Besides a spoken intervention, there was also a musical one. The group sang a song that we learned at Canticle Farm that began, "We must let the earth shape us. Let water carve caverns within us."

After the scroll was unfurled, youth activists took turns speaking. Orion sought to create distance between their community and a Trump presidency. "Trump represents hatred, climate denial, sexism, and forces of oppression that we do not identify with at all."

The activists present vowed to continue fighting for climate justice, no matter who was in office. A ring of press people with cameras and microphones looked on.

"We are all connected," Orion sang.

Becky Chung spoke at the Global Campaign to Demand Climate Justice press conference. Her voice reached readers of Reuters,[4] Deutsche Welle,[5] *El País*,[6] and Breitbart[7]: "I'm eighteen years old and voted for the first time for Hillary Clinton a few days before I left for COP. My heart is absolutely broken at the election of Trump. I have a president who no longer values who I am as a young woman or the child of immigrants," she said. "Our country must undergo a systemic change and just transition away from fossil fuels toward renewable energy within my lifetime. The next four years are critical for getting on the right pathway."

No matter what world leaders decide, the message was that youth will be on the front lines pushing for more decisive action on climate injustice.

David Tong of New Zealand led the Aotearoa Youth Leadership Institute's delegation at COP22. He's a veteran of five COPs and doesn't mince words. "We learned in Copenhagen, if not before, that the United Nations will not save the world," David told me. "We should damn well know by now that governments will not save the world. It's people that save the world. Governments just tend to get in the way."

A central irony of COP22 is that people with the most at stake in the negotiations—youth, women, Indigenous people, people of color, and residents of the global South—were largely excluded from negotiation spaces.

"Youth representation inside COP is very circumscribed," David said. "Civil society doesn't have much it can do." The only time international youth are allowed to sit in the negotiations is when they are giving an intervention: a two-minute speech at the end of a long negotiation session. As a group, international youth are allowed just two or three interventions at COP. At the end of two minutes, the microphone is switched off.

"Back in Durban," David recalled, "my friend Anjali Appadurai gave an intervention. At the end, a party member said, 'I wonder why it is that we let youth speak last, not first.'" In Lima, Tong watched Kya Lal, the only youth representative from the Pacific Islands at the negotiations, deliver an intervention—a deeply personal story about her home islands—to an empty room after the negotiations had finished.

In a time when negotiators are questioning the point of gathering every year to set ambitious targets that many countries aren't meeting, Beatriz offered perhaps the most compelling reason for COP to exist: so that global activists can connect, exchange organizing skills, and take those strategies back home.

"Even if we fail, and we get a 3.5- or 4.5-degree world, I can't think of a bunch of better young people to fail to save the world with," David said.

MUSIC

At the end of the first week, SustainUS reserved a space inside one of the civil society tents. We were all tense. Music, we reasoned, would help.

Dineen played guitar. Daniel joined in on the mandolin. Ben brought his violin. Some of us sang. A few dozen conference attendees stopped to listen. Then, the crowd swelled into something larger. People started dancing and tapping their feet. The delegation kept singing, until a security guard asked us to put the instruments away. People began referring to us as "the SustainUS Choir."

What I learned from interacting with activists is this: show up. Show up and listen. Listen and amplify. Maybe this was something I knew already, but it was concretized in the company of organizers.

"At the core of storytelling is first story-listening," Brooke said, reflecting on the event four years later.

"Listening is at the center of relationships," Kailea added.

SustainUS built community through song and dance in a space that felt so devoid of emotional connection. Sharing stories helped us bridge gaps across our varied experiences.

"Keep play in your body—that's just as much an ingredient in being well and making effective work happen," Niria remembered. "We did sing a lot."

There is always, if you listen closely, a moment of beauty in chaos.

BICYCLES

The conference, while staid from a distance, was full of characters. One morning, in line for coffee, I stood next to a man wearing a yellow bicycle helmet, chinstrap buckled. From the chin down, he was business casual, but his head was encased in neon. People in suits surrounded us. There were no bicycles in sight.

He greeted me with a smile, a pleasant surprise. Inside COP, eye-to-eye contact with strangers is rare. Open smiles are almost unheard-of.

"I have to ask about the bike helmet," I said.

He introduced himself as Andy Costa, an ecologist and cycling advocate from Côte d'Ivoire. We stepped out of line to hear each other more clearly, standing side by side under a palm tree.

"I'm wearing this bike helmet for the future of Africa," Andy told me. "My presence at COP22 is to speak to Africans who face global warming."

For the past three years, Andy had worked to promote cycling culture in his home country. His project, Green Transportation for All, involves installing bike stations in universities across Côte d'Ivoire.[8] He hoped to extend this work elsewhere in Africa.

"Students are the leaders of tomorrow," Andy told me. "If we train and immerse students in a green culture, then they'll continue to live green."

Back home in Côte d'Ivoire, Andy gives presentations in front of the National Assembly and in public to encourage the country to think of cycling and development as going hand in hand.

"Right now we don't have much in the way of cycling culture," he told me. "In the general population, people know how to ride a bike, but don't use it much."

Andy's goals are threefold: to give young people a green method of transport, to promote health, and to increase mobility.

"My presence here represents a change in mentality," Andy said. "We need to change how we think about development. The last time President Obama visited Africa, he said that Africa is the future. Africa can change the global conversation."

In a country where the majority of the population is under twenty-five, youth hold the key to cultural change. Andy, with his focus on university students in Côte d'Ivoire, seemed to understand this clearly.

"Children are the most important part of the future. When you go to NYC or Copenhagen or Paris, people get around by bicycle," he said. "Bicycles are cultural. When I go home to Côte d'Ivoire, people tell me I'm crazy. And that's okay, because it's the good craziness that will change the world."

Here's to more of that "good kind of crazy"—goodness knows we need it.

COP22 also marked the launch of Medina Bike, Africa's first bike-share scheme.[9] After COP, the bicycles continued to exist in major urban points in Marrakesh, with the intention of spreading to other African cities. The goal of Medina Bikes is to alleviate urban congestion and pollution in a city center of tangled traffic.

BHUTAN

I connected with Thinley Namgyel, a negotiator from Bhutan, via a mutual friend. We met outside the conference tents in the early days of the negotiations. Thinley told me that he had attended COP every year since the Bali conference in 2007. In that time, "the mood has really changed," he said. Since Paris, it has become more positive. He sees the Paris Agreement as an achievement. "Now the mood is: How do we implement Paris and then get the action on the ground?"

The main environmental issues in Bhutan, Thinley said, are connected to economic development, waste management, and urbanization. While the country has about 60 percent forest cover, strong water quality, and intact biodiversity, he worries about air quality in cities with a growing number of vehicles.

"What's coming from outside, the threats from climate change," Thinley said, are even bigger than the threats of domestic pollution.

"We're seeing the warming effects," he added. Migratory birds have moved to different altitudes. Rainfall patterns are changing. The timing of the "flowering of the seasons," Thinley told me, has also shifted. Snow is less frequent in the mountains. Glaciers have melted and receded.

"We have very visible impacts of climate change and we're quite worried about what might happen, particularly with water resources," Thinley told me. "A lot of the communities depend on water coming from springs. Those seem to be drying out. Our forest cover is still pretty good, so we're not sure what's causing it, whether it's some geological things or climate, we need to figure it out. But we think climate change has a lot to do with it as well."

I asked Thinley what he has learned from attending so many UN climate conferences.

"Despite the huge differences, collaborative action is possible," he said. "When I came to my first COP, it was overwhelming. I didn't know where to go. There were multiple meetings going all over the place. I couldn't follow the process. Now it's become even more complicated. But somehow we managed to come up in the end on a consensus," Thinley added. "And in this process, I think consensus is important."

In a multilateral process, conversations can get messy. "It's not just countries, but also now nongovernmental actors like the civil society businesses. Everybody has to be involved in this," he said. "It's slow, cumbersome, but we need everybody to work on this." Thinley told me that he is optimistic that countries will be able to honor the Paris Agreement.

His role as a new father motivates him to keep going. "I've always been interested in the natural environment. But now, as a father of young children, I feel that it's imperative that we do something for the next generation," Thinley said. "When I was younger, I just felt it was the right thing to do because we had a close connection with nature back home in Bhutan and we see it everywhere." But as a parent, Thinley added, he also wants his children to have a better future.

UGANDA

I talked to Gertrude Kabusimbi Kenyangi, a Ugandan forestry activist, in the back of one of the civil society tents. She noticed my cardboard sign, made eye contact, and offered to tell me a story about climate change. In 2015, Gertrude won the Wangari Maathai Forest Champions Award in recognition of her work to promote the conservation and sustainable use of Uganda's forests.[10]

Gertrude comes from Katojo Village, a community in southwestern Uganda where the major economic activity is small-scale agriculture.[11] "We don't have a lot of advanced technology," she told me. "We still use the hand hoe. Our agriculture is rain fed and we are using our indigenous seeds. We don't use a lot of pesticides or other agricultural chemicals, not out of choice," she added. It is simply what is affordable and available locally: "organic agriculture by default." Bananas, peanuts, kidney beans, and peas are staples.

Gertrude traveled to COP22 in Morocco to advocate for gender equality in climate policies and solutions. "Those who are most affected by climate change and are not able to adapt, they cannot absorb the shock. And it's almost impossible for them to bounce back from any climate disaster," she added. "Women in particular are very vulnerable, especially in countries in the global south."

In Uganda, climate change has impacted agriculture. A prolonged drought in 2016 caused crops to fail. On the opposite end of the spectrum, violent rains have prompted flooding and landslides. "At one time, there was a landslide that claimed three hundred lives, women, men, and children. The head count revealed that the majority of those whose lives were claimed were women," Gertrude told me.

She attended COP19 in Poland and COP20 in Peru, but she couldn't find funding to support travel to COP21 in Paris. "I feel humbled that I have the opportunity to come and represent a large population," she said. "Working in isolation, we are not going to achieve much." For Gertrude, collaboration is an essential part of the conference. "Creating synergies is our most important strategy."

Her hope for the negotiations in Marrakesh was that "whatever policy positions are taken here can translate directly into practices, because usually policy and practice are divergent. Policy positions are usually good. But when it comes to putting them in practice, there are lots of other forces," she added. "Many policies end up on the shelf."

She also hoped "that the powerful nations of the world would stop being selfish, and would adopt methods that are more sustainable and transformative, not just for the sake of least developed countries, but for themselves as well," she said. "We are all in this together."

"When you think about Uganda specifically, what does climate change there sound like or smell like or taste like?" I asked.

"It tastes like a shortage of food. It tastes like shortage of water because our agriculture is rain fed and we are dependent on natural sources of water," she told me. "So when the aquifers dry up and our crops fail, it smells like disaster. It's so evident that climate change is taking place. The seasons have shifted. When we expect rain, it doesn't come. And when it comes, it's too short a duration; you cannot plant anything. So we don't have any fallback position. We don't have reserves of food. We don't have money stashed in the bank. We live day by day."

BOLIVIA

Carmen Capriles is an ecologist and environmentalist from La Paz, Bolivia. She traveled to COP with the Women's Earth and Climate Action Network to represent her local organization: Reacción Climática. We found ourselves in many of the same side events during the conference. I asked Carmen if she had a story to share. We gathered around a wooden table near a tent selling food to discuss Indigenous sovereignty and environmental issues.

In Bolivia, Carmen works on Salvemos al Madidi, a campaign that has been ongoing since 2004. Madidi National Park covers more than seven thousand square miles ranging from lowlands to nearly twenty thousand feet above sea level.[12] Oil drilling, road construction, mining, new human settlements, plans for megadams, and deforestation for soybean and meat production all threaten the park's careful balance of biodiversity.

"We're talking about one of the hot spots of biodiversity," Carmen told me. The park includes wetlands, cloud forests, lowland jungle, rivers, streams, and glaciers. A 2015 study, Identidad Madidi, contributed to the tally of 8,524 species within the park's boundaries. There are also Indigenous communities who live there: Tacana, Uchupiamonas, Lecos Apolo, and Lecos Larecaja. Some of them do not have contact with the outside world.

In La Paz, Carmen told me, "we listen to their demands and we try to help." When there was a flood in the Beni River watershed in 2014, Reacción Climática helped Indigenous communities living there to get access to drinking water. Another request came for papaya seeds. "The next year we were able to go and see the trees, how they have grown," Carmen said. "That was really encouraging."

In addition to fulfilling requests for the communities, Carmen facilitates conversations about climate change. "In the urban areas we are very disconnected," Carmen said, but Indigenous communities often experience the impacts of climate change firsthand. They use information about weather, the timing of the seasons, and the potential for floods to make decisions about when to harvest or sow seeds.

Reacción Climática has been able to work alongside Indigenous communities to prepare for the climate future. "If you're going to plant land, let's say near the river, you have to be aware that it may flood any time," Carmen said. Instead, the idea is to plant in another area so that the seeds will be safe from flooding. Resiliency is key.

A megadam, however, is another matter entirely. Studies are ongoing for two hydroelectric projects inside Madidi National Park: the Chepete and the Bala.[13] Creating these dams would flood a thousand square kilometers of tropical rainforests and jungle. The dam is a transition corridor for fish species who swim upstream to lay their eggs. "All those systems are going to be destroyed," Carmen told me, if the dam construction goes forward.

Several Indigenous nations call this territory home. In an area largely without roads, the river is a lifeline for communication and transportation. Each community has their own language and their own ways of understanding nature. There is no single opinion on which path forward is best. "We cannot force everybody to develop at the same time," Carmen said. "Not everybody wants or needs a cell phone, let's say. But you have to give them the opportunity to at least choose."

If the area is flooded for the dam, plant matter will degrade under the water and emit methane into the atmosphere. "We know by now that methane, in the short term, is much more dangerous than CO_2," Carmen told me. The dam project would turn a "beautiful place into a big swamp." Communities would be displaced to nearby cities. "And since they are not prepared for living in the cities, they're probably going to engage with the poverty belt around the cities," she said.

From a gender perspective, displacement "is much harder for women," Carmen added. "Many of these women have never been out of the community. So having to face the fact that they're going to lose not only their houses but their whole communities—it's a very big burden for them," she said. "The despair that they have, it's indescribable."

There is a small group of local people who oppose the dams, but they don't have access to international spaces like COP. Carmen hoped to build on support from the Women and Gender Constituency, a stakeholder group of more than thirty civil society organizations that works to ensure that women's voices and their rights are central to the climate negotiations.

"The authorities should know that the megadams are not good policies because they are expensive. They destroy the natural habitats, they influence local communities, and the benefit is not high." Carmen supports solar energy which, in her estimation, is more democratic.

ROOFTOP

After the negotiations ended, the SustainUS delegation left Marrakesh to reflect. We traveled to Imlil, a town about sixty kilometers south, in the high foothills of the Atlas Mountains. We were surrounded by orchards. In the afternoon light, we walked to the top of a hill overlooking the harvest. The whole town smelled like apples in the sun.

At night, after warming ourselves by a fire, we made our way up to the roof of the guesthouse. We bundled ourselves in blankets and looked up. A meteor fell.

"You see that?" someone shouted, pointing. We hushed them. Orion wanted to tell a story. It was like being in a sound booth, but quiet and cold.

"My story starts out first trying to link together issues of water and climate change with societal struggles and dysfunctions," Orion said. Orion, who uses the pronoun they, hails from Stockton, a city in central California alongside one of the largest rivers in California, the San Joaquin.

Stockton was hard hit by the financial crisis in 2008. In 2012, the city went bankrupt.[14] In addition to coping with evictions, gang culture, and drug abuse, Stockton is consistently ranked one of the least literate cities in the nation, Orion explained.

Orion started out in community organizing. They applied for a grant with the city art commission to put on small-scale benefit shows and concerts. Half of the proceeds went to nonprofit organizations, and half toward continuing the events. During these two years of organizing, Orion learned about water privatization.

Now, their activism is rooted in the San Joaquin River, "one of the most crucial water resources, arguably, in California," and a source for agribusiness operations. The river flows into the Sacramento–San Joaquin Delta, the largest estuary on the West Coast.[15]

Starting in 2010, Orion worked as an organizer with Restore the Delta, a group trying to protect waterways from further development. In the 1980s, a diversion project called the Peripheral Canal went up for a public vote. Californians decided not to implement the project.

"The project kind of reverberated into this present incarnation, which was originally called the Bay Delta Conservation Plan, which was a very greenwashed name for extracting the rivers dry," Orion told me. The project proposes to build two thirty-five-mile-long tunnels that would pump water away from the Sacramento–San Joaquin Delta.

"We've been pushing to stop this for a long time," Orion said. "When I started doing organizing work, I first thought that the answer was cross-cultural harmony and intergenerational communication." Then, they realized that the issues were interconnected, "that corporate intrusion and privatization of resources was an actual problem."

And slowly, Orion saw things change in Stockton. A "groundswell of different people" started taking small actions, whether it was a small business owner who put on a benefit and donated money to the homeless, or the university putting on a TEDx event to bring inspiring thoughts into the community.

"I think that in periods of immense stress and crisis, there are always these beautiful, resilient, solution-based tendencies that arise from people in the wake of immense challenges," Orion said. "What gives me hope is understanding the duality of human experience and that people rise from even the most unlikely of circumstances."

Orion noted that as a human, it's easy to feel helpless, especially when the system is so much larger than any one individual.

"One thing that gives me a sense of renewed energy in doing this work," Orion added, "is that what humanity has been able to do, although horrible at its end, also shows the immense power that human beings have in terms of making impacts."

They took a breath before continuing. "It just happens to be the wrong impacts," Orion continued. "I hold on to the hope that all of us have that capacity collectively to change the world."

The stars were a carpet above us. My legs were so cold I could barely feel them. In someone else's story, I felt, I could travel anywhere.

13

UNITED STATES

DISTRACTION

Minnie Mouse tapped Mickey's shoulder. "*Mira, la chica con el letrero,*" she mouthed, pointing to the cardboard sign hanging around my neck.

Mickey sauntered toward me in his big, red shoes. His oversize mouse head obscured any view of the eyes within. I took a guess at where to look as we spoke.

"You want a story about climate change?" Mickey asked, a note of incredulity in his voice. His words were muffled slightly through the headpiece, and his voice was deeper than I imagined. I nodded and pointed at the audio recorder in my hand.

"Just you wait," Mickey said. "This whole thing—Broadway, the lights—is all an illusion. Deep down, something is very, very wrong."

Mickey gestured at the block of Times Square around us. "It's a distraction. All this Hollywood, everything. Business as usual. You would never know what time it is," he said. "There's a lot more you don't know."

Though it was after ten p.m., it felt like daylight with all the illuminated advertisements hovering over our heads like gods: Janelle Monáe and Google Play. Gillette Razors and Coca-Cola. A beacon of light, this America.

"The water's gonna come and wash all this away," Mickey continued. "The whole coastline of the country will change, this year, even. When it happens, it will set our country back a hundred years. There will be wars over water. Just you wait.

"We put ourselves too high, too mighty. It's going to bring us back to earth."

I glanced away from Mickey's plastic stare to watch a teetering child and his two French-speaking parents approach Batman. The father tried to hand his child over to the unsmiling character for a photo. The boy started to wail.

THE DINNER PARTY

On the fourth floor of the Brooklyn Museum, there are thirty-nine plates arranged around a triangular table, thirteen on each side. Each place setting is embroidered in gold thread with the name of the woman who should sit there: Virginia Woolf, Sappho, Sojourner Truth, Hypatia, Artemisia Gentileschi, Sacajawea, Georgia O'Keeffe. The plates are individually painted in oranges and purples and blues, and vulvic by design. Flower petals and liplike ridges fold on the china-painted surfaces.

This is a 1974–79 piece by Judy Chicago: *The Dinner Party*.[1] In September 2014, while walking around the perimeter of the triangular table, I met Mollie Burke, an artist and a Vermont state representative.

"It's appropriate to be here right now," Mollie told me. "The state of the world is a result of women's voices not being heard and a male-dominated culture of exploitation of resources."

It was the day before the People's Climate March. Mollie had traveled from Vermont to New York City "to show world leaders that really the time is now, the time is way past, to start majorly cutting emissions." We both had the same idea—that paying homage to *The Dinner Party* would be a good place to start.

Mollie started to be concerned about climate change in the early 1990s, when she noticed that Vermont's winters were changing. At first, she thought that the changes in snowfall were "just an aberration." She convinced herself that climate change was something far away—a distant future. "I had this uneasy feeling because I love winter," Mollie told me.

She said that not only does climate change threaten her personal world, but also the Vermont economy, "which is based on foliage viewing, which needs cold nights to produce the vibrant foliage." Maple trees also need a particular kind of temperature variation in order to produce sap for maple syrup. And the ski industry depends on snow.

"So much money has gone toward climate denial—which is really very sad—in this country. Our democracy has been usurped by moneyed interests and greed," she added.

"It's a very crucial moment in history right now," Mollie said. "I think it's really important not to feel depressed and devastated. And that's why we're here in New York, to stand witness to these feelings."

THERE ARE NO STRAIGHT LINES

Sound comes to our bodies in the form of a wave, entering the outer ear and traveling through the narrow ear canal to the eardrum. When the eardrum meets these waves, it vibrates and passes the movement to the three tiny bones that compose the inner ear: the malleus, the incus, and the stapes. These bones amplify the sound waves and send them to the inner ear's cochlea, which looks like a fluid-filled snail. The fluid ripples, activating sensory cells on top of the membrane—hair cells—that ride the wave. The tilting of this motion causes the surface of the bristly structures of hair cells to open, allowing chemicals to enter that create an electrical signal. The auditory nerve carries this signal to the brain. The brain translates the current into a sound that is recognizable. And so I pass through my day: listening, fluid.

The sound was dense, thick: a river of human voices. With headphones over my ears, the world was amplified. There was the rumble of a gathering drum ensemble to my left, the growing pulse of conversation, heartfelt hellos, and a rising song—movement everywhere.

On September 21, 2014, four hundred thousand climate activists had gathered in New York City for the People's Climate March, an eighty-five-block-long tide of humanity walking from Central Park to the United Nations—at that time, the largest climate march in history. The purpose? To demand that world leaders take meaningful measures to address the climate crisis. I was a drop in the ocean.

Some people flew. Others took a train from California. Many arrived by busload from Pennsylvania and New Jersey and Rhode Island. I took the train from Connecticut to Manhattan and crashed on a friend's couch in Brooklyn. The morning of the march, I woke up early and put a reinforcement layer of tape on the cardboard sign I wear around my neck that says "tell me a story about water" on one side and "tell me a story about climate change" on the other—the edges where the ribbon connects with cardboard sometimes came loose.

The subways were packed to inches. I exited the train at Times Square and walked the remaining blocks to Central Park, following a woman who looked like she knew where she was going. She carried a sign that read "System change, not climate change" in blue Sharpie. Her boots clicked on the sidewalk.

"Are you going to the Climate March?" I asked, stating the obvious.

"Of course." She glanced at me sideways, not pausing in the brisk rhythm of her steps.

"Where are you from?" I struggled to maintain eye contact while weaving in and out of the gathering mass of people. New York City always feels full, a pressing on my temples. The city throws into sharp relief the fact that my parents are mountaineers, that I was raised in the woods. I come from trees and streams and backyard tire swings, from bike rides on empty suburban roads. A train connects my hometown to the artery of New York City, but I grew up looking north to the mountains. My family coveted wild places, the uninhabited ones. The only times I braved the trip to the city were with theatergoing friends and their cosmopolitan parents who worked in finance and generously funded our Broadway tickets to see *Rent* and *Wicked*. My long weekends growing up consisted of family camping trips and long walks over crunchy leaves.

The woman marched on. "I'm from across the river in New Jersey, but I grew up a New Yorker."

"You sure do walk quickly," I said, struggling to keep pace.

"It never really leaves you, the place where you come from."

I lost her voice in the collective pulse of drums.

"We are not sinking, we are fighting!"

Pacific Climate Warriors led the march, followed by other Indigenous groups at the front lines. "Front lines of crisis, forefront of change," their banners read. The crowd was a rotating kaleidoscope: big splotches of yellow, triangles of red and green. Alaskans alongside Pacific Islanders alongside New Yorkers. Far above us, the lone eye of a drone camera recorded. Years later, what will be told of this story? Whose eyes and ears remember, and how?

I wore my cardboard sign as I ducked and wove through the crowd backward, moving against the upwelling of activists. Sometimes I just stood still and let the swell of voices come to me. I tried to intersect with as many storytellers as I could.

"Climate change is a health crisis," said a man in a white lab coat.

"Are you part of government surveillance?" someone asked me.

"One of the most graphic things was a story someone told me about a ninety-year-old in the Arctic who said 'I've never seen these birds before.' Birdwatching has picked up in the far North," another person said.

A man balanced an oversize inflatable globe on his shoulders. I passed a row of human sunflowers and a woman wearing a beekeeping suit and holding a sign that said, "RU Living in a Bubble?" A young mother with an infant wrapped on her front wore a placard that read: "My future is in your hands." She breastfed amid the chaos.

The color of this movement—big gestures, big words, big buildings framing our motion—is something I still carry under my eyelids. Even when I blinked, I could feel the enormity of the crowd, stretching for blocks in every direction. The streetlights arched above us, lights changing from green to yellow to red. For hours there were no cars, only people.

"What do we want? Climate justice. When do we want it? *Now.*"

"Our future, our choice. Let's use our voice."

"El pueblo unido jamás será vencido. El pueblo unido jamás será vencido."

"Are you, like, a reporter?"

"Free bagels? We have to keep ourselves fueled today. Here, I'll tear one in half with you."

"The world is ours. The water is ours. The air is ours."

"I say climate, you say justice. Climate . . ."

"Justice!"

"Climate . . ."

"Justice!"

A marching band amplified exuberant brass. The trees of Central Park listened on. A helicopter passed, momentarily covering the sound of everything from above. I looked up to the soft gray patch of air between the buildings until another storyteller brought my attention down to street level.

"My name is Joyce. I'm from Arkansas. I am seventy-one years old. I feel like my generation has messed things up so bad that I had to be here today, and I hope that the young people understand that they have a real huge job ahead to make the change, but if there's any hope at all, it's in our young people. Good luck."

At noon the crowd fell silent. A hush simmered. Above us, people hung out of their apartment windows, watching. The seconds stretched—elastic, pulsing.

Then, from blocks behind, a tidal wave of sound emerged. Before I knew what was happening, I was caught up in the noise, shouting and stomping my feet and jumping, jumping in unison with the hugeness of the crowd around me. It was a free fall of voices. Overwhelmed, I cried. I had never been a part of something so large.

I saw a sign to my left: "It takes everyone to change everything."

The person standing next to me turned and said: "So are you ready to hear my story?"

SANDY

Ken Gale is the host and producer of EcoLogic, an environmental radio program at WBAI FM. When Hurricane Sandy hit New York City on October 29, 2012, he was on the air. We met at the NYC People's Climate March in 2014.

"We used to be on 120 Wall Street, on the tenth floor. We knew Sandy was coming, and in order to keep the radio station running we needed a skeleton crew, and I was part of that," he told me.

The building had a lobby area with a window overlooking the East River. Throughout the day, Ken and his colleagues stopped at the window to monitor the storm. "At one point water started pouring across South Street. And then it would hit the sidewalks on the corner, and then go up Wall Street and then pour over the edge of the sidewalk and then go up higher and higher," he said.

The program director let the team know that they would soon go off the air. They told the listeners that this was it. "We had one minute and fifty-eight seconds, and all the equipment went off with a big loud crack. It sounded like a gunshot, all the electronic equipment shutting off at once when the power went out," Ken said.

He went back to the window to watch the water come higher. Seawater started pouring into a fountain in the front of the building. Then a storm surge brought water above the traffic lights. Ken and his colleagues were stuck in the building for sixty-two hours. After the water receded, somebody with a car came to pick them up.

For the first few hours, there were emergency lights on the wall. Later those went out, too. "We were totally in the dark. So it's very bizarre, you know, feeling your way to the bathroom that you've gone to so many times but now it's in the dark," Ken said.

"People ask me if I was scared, that's the number one question," he added. "No, because I know what a hurricane is. I know what a storm surge is. I know what tides are. So I knew what was happening. And I knew it wasn't going to get ten stories up." He watched, fascinated, at the window. In the middle of the night, the storm surge was gone. "Low tide took all of the water away."

AVOCADO

I met Rebecca and her daughter Athena at the Climate March, too. At the time, Athena was five months old; she slept on her mother's chest as we walked. Rebecca told me that she and her husband moved to Ojai, California, because they wanted to start a family and get out of Los Angeles.

"We moved to a five-acre avocado ranch in Ojai, and we have three-hundred-plus avocado trees," Rebecca said. "Our whole goal is to build and create this farm so that we could raise our daughter, Athena."

Rebecca wanted to create a farm that could be passed down through generations. A prolonged drought made that goal more complex than she had anticipated. "We're now at the end of a five-year drought and our water is running out, and so our trees are extremely thirsty," she said. "The cost of water is soaring. We're not sure if we're going to be able to keep our trees or not because of how water-consumptive it is."

Most of the water in Ojai comes from Lake Casitas. After five years of drought, the reservoir was on its last reserves. "The climate, just since we've moved to Ojai, has completely changed," Rebecca said, "and so our big dream and our vision of moving to this beautiful place where we could grow our own food and we could live sustainably on the earth is starting to die, quite literally. It's drying up."

LOSS

Three years later, I attended the 2017 People's Climate March in Washington, DC. Melissa Hawthorne saw my cardboard sign just as we were passing in front of Trump International Hotel. We tried to find a quiet space to record as people streamed past us, chanting "Water is life!" and "Shame! Shame! Shame!" at an oversize papier-mâché Trump head. The president had announced his intention to withdraw from the Paris Climate Agreement earlier that year.

Melissa was thirty-five years old and had lived in Florida all her life. She went to the Florida Keys for the first time to snorkel when she was fifteen. "It was so beautiful," she said. "It was just like you would imagine from *National Geographic*. Every color, every shape of coral. Pink, purple, blue," she smiled. "All kinds of fish and sharks."

As she spoke, it was clear that the coral reef holds a specific power in the landscape and waterscape of her memory. The Keys are linked to her childhood—a sense of beauty and awe in the underwater world that shaped her.

Melissa returned to the Florida Keys two weeks earlier to snorkel again. "The reef is 95 percent dead," she said. "When we got into the water, there was nothing to see."

"How does that feel?" I asked.

"Horrible," her voice broke. "I saw for the first time in my life that my children won't get to experience or see the same world that I do. Thank you for letting me share my story." She continued to march in tears.

There's a word for the kind of grief that Melissa expressed: "solastalgia." This term describes the feeling of loss and distress caused by environmental change—an emotional disquiet from witnessing negative changes to one's home. The term was coined by Australian environmental philosopher Glenn Albrecht in the early 2000s.[2]

The ocean of Melissa's childhood has changed for the worse. Those colors and shapes of coral now live only in her memory.

AQUANAUT

Grace Young is an aquanaut and ocean engineer. In 2014, she lived sixty feet underwater for two weeks in a habitat called Aquarius off the coast of the Florida Keys, the last working underwater lab in the world. The habitat is the size of a yellow school bus and housed six researchers at the time. Inside was like an RV and a space station, with six bunk beds, a tiny kitchen, and freeze-dried food. Outside the viewports from the kitchen table, Grace saw eagle rays, turtles, and goliath groupers.

"I would do it again in a heartbeat," she told me. On a clear day, she could see the outline of the sun in the water above. At night, it was pitch black.

While living in Aquarius, Grace and her colleagues studied how the reef around the habitat has been changing. One of the advantages of an underwater lab is that it lets scientists spend more time up close with their subjects. A diver can only spend up to forty minutes at depth before returning to the surface for a maximum of three dives a day. The underwater station enables saturation diving, which allows divers to stay at depth for extended periods of time. They monitored how plankton, the base of the ocean ecosystem, changes between day and night. Grace also looked at temperature changes on different points in the reef, and contaminants in the water column.

"The deep ocean is still a giant mystery to us," Grace said. "The tragedy is that that ecosystem is slowly degrading, even disappearing, before we even know what's going on or even that some ecosystems existed." Climate change, pollution, and irresponsible fishing practices are all to blame.

GROCERY

We met by the hamburger buns at the Kroger grocery store in Clarksdale, Mississippi in 2014. She was riding a motorized shopping cart. I had just put my audio recorder away after speaking with a man who was stacking apples.

"Tell me a story about climate change," she read aloud from the cardboard sign around my neck. "You know, I feel like I can talk to you."

I took a step closer to listen.

"We don't care enough about nature," she began. "We keep taking and taking and not giving anything back. I'll tell you what. It's hotter now than it used to be. I remember cooler Octobers in Mississippi."

She shifted in her wheelchair, eyebrow ring glinting under the fluorescent lights.

"Some people call me Hippopotamus. They don't want to listen to me because I'm big. But you're listening," she said.

"I tried to kill myself nine times. The last time I shot myself and had to go to the hospital. They asked me if I hear voices. I said, 'Doesn't everybody?' "

She took a deep breath. "I hear the trees. I hear when they're angry. Yesterday I heard an angry wind in the trees. The trees talk to me. They comfort me."

I nodded and continued to listen.

"I can't believe I'm telling you this. You're the first person I've told this to. I can't talk to the people around me because they'll lock me up. All they want to talk about is who's fucking who and Prada and Gucci. I don't want any of that. My grandmother was a midwife. She was the last one who understood."

The woman exhaled, long and slow. "I feel so alone. I've been to the dark places. The depression can get so bad."

Sometimes, people just want to be heard.

DOOM

Charlie's friends call him Yoda. When we met in New Orleans in 2014, he told me that he works as an equine therapist. "Normally I'd be traveling about, working on horses," he explained. "I'm stuck right now because I'm battling cancer."

Charlie told me that New Orleans sits below sea level. "You have funky water. You have funky energy. It's a place where you can feel okay about being funky," he said.

"When you look at the elements of climate change, we've polluted our world so bad," Charlie told me, "because we're greedy. We're pigs. Our society sucks. Not only America, but everywhere. It's all about one-upmanship and what can I do? What can you do for me? Politicians, they're a disease. A bad disease."

Behind us, the St. Louis Cathedral bells rang the quarter hour.

"What can we do? Nothing. The machine is in motion. We're doomed. So the best thing you can do is hop on the best bus and enjoy the ride. Do the best you can do," Charlie said, reflecting. "Like me, you know? They gave me two years to live. So I'm going to live for two years. C'est la vie."

At the end of his life, Charlie made a zen garden in the backyard of his family home. "I never knew I had a green thumb, man. I put these plants together, you know, and it's okay," he said, smiling. "I'm going to dance, have fun."

DENIAL

The United States is not neatly divided into climate change believers and climate change deniers. In November 2017, I attended the Yale Environmental Sustainability Summit, where I heard Anthony Leiserowitz, director of the Yale Program on Climate Change Communication, discuss his team's research. Since 2008, the program has been conducting surveys that track how audiences within the American public respond to climate change in distinct ways based on their beliefs, attitudes, policy preferences, motivations, risk perceptions, values, and underlying barriers to action. They identified six groups that they call "Global Warming's Six Americas."[3]

The *Alarmed* are convinced that global warming is an urgent threat that is caused by humans. While they strongly support climate policies, most in this group do not know exactly what they can do to solve the problem.

The *Concerned* consider global warming to be a serious threat and support climate policies. That said, they believe that climate impacts are still distant in time and space, which makes climate change a low priority compared to other issues.

The *Cautious* haven't yet made up their minds on whether global warming is happening, is caused by humans, or is serious.

The *Disengaged* know little about global warming and rarely or never hear about it in their consumption of media.

The *Doubtful* don't think that global warming is happening or believe that it is part of a natural cycle. They don't consider it a serious risk or think much about the issue.

The *Dismissive* don't believe that global warming is happening, human-caused, or a threat. Many in this group endorse conspiracy theories—that climate change is somehow a hoax perpetuated by scientists or the liberal left.

As a whole, Americans are becoming more worried about global warming, more engaged with the issue, and more supportive of climate solutions. From 2015 through 2020, the Alarmed segment of the US adult population has more than doubled in size (from 11 percent to 26 percent), while the Dismissive segment has decreased by nearly half (from 12 percent to 7 percent). The Cautious, Doubtful, and Dismissive groups have been shrinking in recent years.[4]

We are starting to wake up.

PRECIOUS

I met Jim Stanfield, a retiree, in Noe Valley, San Francisco, on Twenty-Fourth Street. "I had always thought that water would be a precious commodity at some point," he told me. As a kid, all the water he drank was free. Sometimes, when he was thirsty, he took sips from a garden hose.

"No one would even think about buying bottled water, but now it seems that people buy that in preference to drinking from the faucet, although we do have fairly decent quality water here in San Francisco," he said.

When we spoke in 2014, California was in the middle of the worst drought in at least twelve hundred years.[5] Jim noted that the drought in California "has now been attributed to global warming. A new scientific paper out of Stanford models this as a most likely cause of our drought here."[6]

That study, led by Stanford climate scientist Noah Diffenbaugh, combined computer simulations and statistical techniques to demonstrate that atmospheric high pressure in the region, which is strongly linked to unusually low precipitation in California, is more likely to form with modern greenhouse gas concentrations. Climate change, in other words, was the most likely cause.

"We now pay more for bottled water than we do for milk or gasoline per gallon, and I think that just kind of indicates the direction of the scarcity and the preciousness of what we have," Jim said. "We really need to preserve it."

PAPER BOATS

In the late nineteenth century, shipbuilders Elisha Waters & Sons of Troy, New York, mass produced cheap, light, and durable boats by layering sheets of manila paper over a wooden mold.[7] It was an old technology—one ripe for renewal.

"Paper had just started being mass-produced out of wood, so there was a lot more of it. It was seen as this new material that was moldable, flexible, strong, kind of like what plastic is now," Kevin Buckland, an artist and climate organizer, told me.[8]

Kevin was part of a team with 350.org, an international environmental organization, and Mare Liberum—a group of improvised urban boatbuilders, printmakers, adventurers, and gadabouts in Brooklyn—that resurrected the paper boat technology. The idea was to construct a fleet of canoes that could carry their activism downstream.

One of Mare Liberum's goals is to empower people to build their own watercraft. To make each paper canoe, the team used an existing canoe as a mold.[9] They put a stretchy plastic wrap over the vessel. Then they cut thin strips of kraft paper and interlaced seven layers of papier-mâché. When it was dry, they removed the mold, built floorboards out of wood, and then sealed the paper with a layer of varnish. With three people working, the boat could be done in a few days.

In 2014, the team built a flotilla of papier-mâché canoes for a two-week journey down the Hudson River. The paper canoes followed a corridor that transports both fracked gas and crude oil. They floated 160 miles downstream, stopping in thirteen towns along the way to facilitate conversations about the impacts of climate change.

"What was amazing is the folks that showed up weren't climate people. They were river folks in fishing clubs, boating clubs, and residential associations of people who live on the river," Kevin said.

They ended in Manhattan, circled the island, and then joined the People's Climate March.

"We really liked the poetry of the paper boats," Kevin said. To him, it was "trickster logic." He put himself in a vulnerable situation on the river. "People came because of the impossibility of it." Along the way, they only lost one vessel.

For Kevin, traveling on a paper boat was "enacting a metaphor. We are in this time. We are in this fragile moment, on these big currents that are moving. And we have to be very aligned with where we're trying to go and how we're going to flow there with what's going on," he said.

He wanted to remind himself of the importance of water. "I'm trying to reconnect myself with things I've forgotten." As it turns out, traveling on the river was, for him, like being in love. Whenever he wasn't on the water, all he wanted to do was get back out. Kevin learned that "what happens to the water happens to us." The motto of the journey was: "We All Live Downstream."

There are as many ways to be an activist as there are people on the planet.

LOVE

Sometimes, climate change brings people together. In 2011, Tropical Storm Lee and Hurricane Irene spun up the East Coast of the United States, filling the Susquehanna River in central Pennsylvania with record levels of excess water.[10] Retaining walls overflowed.[11] In the city of Wilkes-Barre, the water reached a height of 42.6 feet.

At the time, Amanda Ciampolillo was working with the Federal Emergency Management Agency (FEMA) as a deputy regional environmental officer. "After a disaster like a flood, FEMA has to come in and do work for local communities and we didn't have enough staff. So we put a call out for anyone who lived in the local area to come and apply to work for FEMA. And that's when her résumé landed on my desk," she said.

During the flood, Molly Kaput was finishing up a master's degree in community regional planning. She applied to work with FEMA as part of her grad school internship. Amanda hired Molly, and they realized that they had a spark. "She had a girlfriend at the time," Molly said, laughing, "but we figured out pretty quickly that something was going on." About a year later, Molly was hired in a different department. She transitioned into floodplain management, where she speaks with communities about building codes and the risk of building on a floodplain.

"So we met, got married because of water," Molly said, putting her arm around Amanda's shoulder.

"We would have found ourselves eventually, but, you know, a flood on the Susquehanna River brought us together a little sooner," Amanda added, smiling at her wife.

THE POWER OF THREE

There is magic in repetition. Events happen in threes, and then something changes.

I met Mark Binder at Boston Logan Terminal E in November 2016. He was waiting for a flight to attend the Sneem International Storytelling & Folklore

Festival in Ireland. We both had time to spare, and the cardboard sign around my neck was an invitation to share a story.

Mark told me that a mayor came to a wise woman and said, "Wise woman, wise woman, the water is rising. The town is going to be flooded. What do we do?"

And the wise woman said, "Move the town."

"We can't move the town; this is our home."

"Water goes where water goes," she said.

"I got an idea! We'll build a boardwalk," the mayor said. So they built a boardwalk.

A few years later, the mayor came back to the wise woman. "Wise woman, wise woman, the water's rising. The town is flooded. What do we do?"

"Move the town," she said.

"We can't move the town; we just built a boardwalk. It's a huge tourist attraction!"

"Water goes where water goes."

"I got an idea! We'll build canals."

So they built canals and diverted the water.

A few years later, the mayor came back. "Wise woman, wise woman, the water is rising. The town is going to be flooded. What do we do?"

"Move the town."

"We can't move the town. We just built these canals. The shipping industry's doing great. . . I got an idea! We'll build a dam."

So they built a dam.

A few years later, he came back. "Wise woman, wise woman, the water's rising."

She said, "Move the town."

He said, "All right. We'll move the town."

She looked at him and she said, "Water goes where water goes."

Of course, the water wins. Towns have to move. Mark told me that he had first told this story at a Montessori School on Earth Day. The teachers reached out to him after to say that the students were angry. "I didn't understand why. And it turns out that they didn't believe this story; they thought that people should win all the time," he said. But adaptation can only take a town so far.

MACKEREL

A twenty-five-year-old man approached me at Malcolm X Park in Washington, DC. It was a warm March day in 2017. I was standing inside a blue-and-white balloon sculpture, made to look like water. Julie Zauzmer, a *Washington Post* reporter who moonlights as a balloon twister, designed the booth. We spent a day bending and weaving the balloons together in her apartment. The next day, we carried the balloons to the park and placed them on top of a blue-and-white checkered blanket. This was a place where storytelling could happen.

The man didn't want to tell me his name, but he did want to talk about mackerel. As a kid, he spent the summers at Penobscot Bay in Maine, catching the fish off a dock. "We would eat them immediately. We would sometimes eat them raw because they were so fresh," he told me, the memory of that taste flashing across his face. Sometimes his family would flour and fry the mackerel, the flesh flaky on their tongues.

"I don't know if it's climate change or what, but in the last few years, there have been no more mackerel," he told me, resigned. "We can't catch them anymore."

I followed up on the mackerel with Paul Anderson, executive director of the Maine Center for Coastal Fisheries. "Some of the observed shifts in the ecosystem in the Gulf of Maine are likely due to climate change," he wrote to me in an email. "These probably include habitat and food web dynamics. Mackerel have to eat something, and we know that all the way down the food chain to the plankton is shifting in the Gulf of Maine."

As with many things climate related, there is a level of uncertainty. "It's also true that mackerel, like many other migratory species, have 'good' and 'bad' years, and may cycle on their own," Paul added.

But an observed change is a change nonetheless. From the food web to energy transfers to cycles of good and bad years—we are all interconnected. Sometimes the taste of climate change is an absence. A memory. A lack of flesh.

OIL SPILL

On September 16, 1969, an oil barge called *Florida* hit the rocks off the coast of Cape Cod near West Falmouth, Massachusetts. It was foggy. Fuel oil—189,000 gallons of it—spilled into Buzzards Bay. The wind and the waves pushed this oil onto West Falmouth's beaches and marshes.[12]

At the time, Susan-Marie Stedman was ten years old. She grew up outside Boston and spent summers on Cape Cod, walking around the salt marshes of Buzzards Bay. She loved learning the names of the creatures who lived in the wetland.

The 1969 oil spill, Susan-Marie said, "killed everything in the salt marsh." She remembers "watching it slowly, so slowly come back to life," a process that took twenty years. The salt marsh grasses died. So did the fish. Scientists at the Woods Hole Oceanographic Institution uncovered the oil that lurked in the marsh and subtidal sediments long after it was seemingly invisible from the water and beaches. Species that burrow through the sand encountered a film of oil beneath the surface of the sand as recently as 2010. The *New York Times* reported that the fiddler crabs of Wild Harbor "still act drunk, moving erratically and reacting slowly to predators."[13]

"And that was just a small oil spill," Susan-Marie told me. Others have been far worse.

Witnessing this destruction and renewal galvanized her to want to learn more about science, and how to protect salt marshes. Now she writes policies that protect wetlands for the National Oceanic and Atmospheric Administration.

When we spoke in 2017, within sight of the Washington Monument at the

March for Science in Washington, DC, Susan-Marie was focused on protecting the Clean Water Act. "There have been a number of court decisions that have made it really hard for people who want to do their jobs to implement the Clean Water Act. And so we're working on trying to keep it from being pushed back," she said.

Around us, scientists gathered to participate in the protest. I read signs that said "We have an ocean of problems. Science has solutions," and a yellow umbrella with the handwritten words: "Science rocks."

Witnessing the impacts of one oil spill directed Susan-Marie's whole career. What stories will the wetland scientists of the future tell?

SEVENTEEN

When we spoke in 2017, Laurie was seventeen years old and "really worried about how Chicago isn't Chicago anymore." To her, the weather patterns are to blame.

"We used to have extra hot summers and freezing winters. Now it's just a normal day in the neighborhood," she said. "It's not what it used to be."

We met in front of the Navy Pier Ferris Wheel; it was mid-September and 70 degrees. The sky behind the Ferris wheel was golden. Two days before, it had been 50 degrees. Laurie referenced this temperature swing as evidence of instability.

"I'm worried about Chicago," Laurie reflected, "but hopefully we'll hang in there."

She knew that in her lifetime, climate change could become unmanageable. Laurie told me that she was concerned that the climate would change "quicker than people expected." She couldn't imagine anything farther out than thirty, forty years. And then?

"Hopefully we're alive," she said. "It may be a little quicker than we thought, but hopefully it's just people's imagination, people overthinking, overwhelming themselves. Yes, it is a world change, but it's not the end of the world."

TOMATOES

The one storyteller I saw twice, without planning it, was Trisha Sheehan. We crossed paths at both the 2014 People's Climate March in New York City and the 2017 People's Climate March in Washington, DC. Trisha traveled to the two events with Moms Clean Air Force, a nonprofit dedicated to fighting air pollution and climate change. She told me about the impacts of climate change on her family's farm in New Jersey.

Trisha grew up on what she calls the largest farm in South Jersey. "When I was a child, I remember running through the fields and picking up tomatoes, you know, the Jersey tomatoes, and eating them right off the plant." She would rub the tomato on her shirt to get the dirt off and then take a bite, the juice running down her face and elbows. Her childhood was marked by tomato-stained shirts.

Trisha is the mother of three children. When her oldest was six years old, he was planting seeds on the farm with her dad in the spring. She went to the greenhouse to pick them up at the end of the day. The seed in her son's hand was hot pink.

"What are you planting?" she asked.

Her cousin handed her a bag of seeds treated with fungicide. "In New Jersey, the amount of rain that we're seeing from climate change and extreme weather impacts is causing us to have more rain, which causes the plants to create mold and fungus." The seeds are sprayed with fungicide in order to protect the crops.

Trisha was horrified by the number of chemicals on the plants. Her cousin warned her not to take a deep breath. "And this is supposed to grow into the food that we're eating," she said, alarmed. "My children are not able to have the same experience that I had growing up on the farm only thirty years ago."

THE NEW GODS

When we spoke at the Ragdale Foundation in Lake Forest, Illinois, in 2017, Gus was eight years old. We had just watched an outdoor play called *A Persephone Pageant* made by Walkabout Theater Company. The company featured massive papier-mâché puppets, flower garlands, actors on stilts, and a reimagining of what the new gods in a climate-changed future might do.[14]

"This happened a long time ago," Gus told me, riffing on the spectacle we had just witnessed. "There is this guy named Edu and he was strong." Edu found old cloth from his mom's dresses and he "sewed them all together" to make a rope. "Then he found the biggest cliff," Gus said, "and then he jumped off. And just like he had thought, he floated."

What Edu didn't know was that the cliff went "all the way down to the Earth's core. So he asked for the gods to help." The gods gave him "the strength that he would survive anything." Edu fell "all the way to the inner core, which is made of hot metal." He saw what looked like "a big rock that he could kick, and he kicked it, causing it to spin fast, causing the Earth to break apart."

Then the Earth was broken. "Everyone fell down into the cracks in the Earth." And then there was no one left.

"Since that day, there are still cracks in the Earth that were made by Edu," Gus concluded. Edu's crack "closed up forever." The Earth became his burial chamber.

REWARD

A few years into the 1,001 Stories journey, Karin Stevens, a choreographer in Seattle, reached out to me to ask if I would travel out west to record stories alongside her in the dance studio. Her idea was to use that audio as inspiration for a performance. I said yes. Together we interviewed Lisa Hayward, a biologist in Seattle who worked during the early years of her PhD program as part of a long-term research study in the Arctic in Barrow, Alaska. "You can watch all the documentaries you want about the Arctic, but there's nothing like being there," Lisa told us. "For one thing, you get to smell the cranberry on the tundra. It's just unlike anywhere else in the world."

In an Arctic summer, the sun never sets. Some days, Lisa would continue her fieldwork until two in the morning. At that hour, "things would get kind of weird," she said.

Lisa and her colleagues shared meals with scientists from other groups at Toolik Field Station. One researcher spent up to six weeks alone on an island off the Arctic coast to document changes in seabird breeding patterns—some species, she found, were breeding a month earlier than they had been three decades earlier. "We're seeing these biological changes in response to climate change. That was, for me, something that you couldn't deny," Lisa said.

Returning south, Lisa felt like she had a responsibility to spread the word. She finished her degree and started a postdoc studying spotted owls in Northern California—her dream job. She oversaw a crew that spent six months in the forest watching owls and recording their behavior in response to motorcycle use; they collected owl poop and measured stress hormones. Back in the lab, Lisa analyzed her results. The study was a success. She wrote up her findings for publication and turned in a report to her funders.

"But then it came time for them to ask me for policy recommendations, and I realized I just had no idea," she said. Lisa started the project thinking that motorcycle use is "horribly disruptive" to owls that nest near the trails. But during the course of the study, Lisa engaged with motorcyclists as volunteers to ride past the owl territory. They spent time camping together.

"These people, they're out here with their kids every weekend. They love this forest unlike any other group I've come across. When there was a devastating fire, they were there planting the seed banks," she said. Prohibiting motorcyclists from accessing public land was "probably counter to our larger goals," Lisa said. "We all know that's not the biggest problem that the owl faces."

After this experience, Lisa realized that she didn't want to be a scientist who made graphs and statistics for a journal that "nobody ever reads." Instead, she turned her focus to science communication. Many policy makers don't have a background in science. And many scientists are outfitted with a toolkit of jargon that is inaccessible to people outside the discipline. Lisa wanted to help bridge that gap.

"We've evolved as a species to engage with information that's presented as a story," she told me. "Our brains light up globally when we're presented with information that has this narrative arc to it and has a relatable protagonist and emotional content." As manager of communications for the Northwest Climate Science Center from 2014 to 2017, Lisa connected policy makers with science they could act on. She worked with the Quinault Indian Nation on their plans to move the village of Taholah half a mile uphill to get out of the range of flooding.

For Lisa, hope comes from taking action. "When you can break down a big project into a series of manageable steps, then there's a lot of power in that. In a lot of cases we feel like climate change is this thing that's operating on a timescale that's just not compatible with our brain because the rewards are so far off," she said. "How can we find a way to reward this incremental progress?"

THE MIDWIFE OF DEATH

We sat crammed in the last row on the plane to San Francisco. The woman next to me had a friend who was dying. She said goodbye over the phone, from the most private corner she could find at the Atlanta airport.

I held her hand as we took off. "This is the worst day of my life," she commented, quietly. "I would rather be anywhere else than here. I need to get to my friend before she passes." She looked me in the eye. "At least I got to sit next to a kind stranger."

I fell in and out of a fitful sleep, resting my hand on the woman's back. It was the only thing I could think of that might help.

After landing, the woman turned on her phone and checked her texts with dread. She knew what they would say before she read them: the friend had passed ten minutes before we landed, when we were stowing our tray tables and preparing for the descent.

~

I stepped off the bus at the University of California San Francisco Medical Center about half an hour early to lunch with my friend Eva, a first-year med school student. I found a tree to sit under and took out some words to read. I had all but forgotten the cardboard sign around my neck that said "tell me a story about water" until a woman approached and lifted me from the page with a question—

"What's that all about?"

"Hmm?" I asked, dazed.

"Water stories? I couldn't help but notice that you looked so serene there. I had to ask."

I laughed and told the story of my trip to collect stories. "My name is Sharon," the woman offered. "I'm a Buddhist chaplain at Mt. Zion Cancer Center. I have to go see a client in five minutes," she added, "but come at ten a.m. tomorrow to the fifth floor at Mt. Zion Hospital. I'll meet you there, and I have a story to share."

The next day, Sharon and I sat across from each other in an empty hospital bedroom, the air between us sterile and sharp with disinfectant.

"I am the Midwife of Death," she began, her long, silvery hair catching light from the window. "After seven years of leukemia, my mother was in her last days. I flew out. They decided on cremation. I wanted to bless and anoint her body after death."

Sharon exhaled and smiled. I could feel the feathery weight of her mother's memory in the room, as present as the central air and the PA system.

"My parents had single beds next to each other at home. They held hands when they slept," Sharon said. She didn't want her father to be haunted by her mother's passing. Sharon suggested that they wash her body together. "We have to release her spirit so that you can sleep in this room," Sharon told her father. "The first time I brought it up, he was silent. When I asked a second time, he said yes."

Sharon surged on. "I had my father pick out her nightgown and together we chose beautiful smells," she told me. Sharon's mother passed away at eleven fifteen at night and her bowels released. "We asked the nurse to clean her up.

The nurse left. My dad and I got a basin and we washed her slowly. We were not in a rush. We washed her hair. Dad was so ginger. I kept on slowing it down. I'd take a washcloth and go all the way down her arm. I'd kiss her elbow. I'd trace her freckles. *Hold her hand, Dad. This is this beautiful face that you have loved.* We washed her all the way from top to bottom. We put on lotion. We put on perfume, her new nightgown. Dad was weeping."

"We followed the blessing way, rather than the medical way," Sharon concluded. "A dead body is not remains, not medical waste. A dead body is a human being."

I silenced the audio recorder and let the tears I had been holding back fall freely. Something deep within me unclenched—a substance I associated with the pain of the woman on the plane. We need water to cleanse, to heal, to aid the transition from life to death. There was something else, too, something unnamable, that Sharon's story had loosened.

"Oh, honey," Sharon whispered, reaching deep within her purse for a packet of tissues. "Why are you crying?"

I struggled to get the words out.

"I'm afraid," I managed, looking up. Sharon's warmth urged me onward, to the heart of it. "Afraid of losing those closest to me. It rattles me with fear. I don't know what I would do."

Sharon put her hand on my knee. "Honey, I coach people who are dealing with family members who are dying. That is my job."

The Midwife of Death offered her advice: "You come to a point in the process where the road forks and you have two choices. You can go, *Oh my god, I don't know how to do this*, and completely freak out. You could let yourself be guided by fear." Sharon's calmness was a balm. Slowly my breathing started to settle.

"Or," Sharon said, "you can just show up. Just be there. Be there in love. Say: 'This is a person I loved.' Say: 'I will be present.' My wish for you, Devi—and I would never wish death on anyone—but my wish is that there will be someone there with you in those moments who has dealt with it before. A leader to follow."

I looked around at the empty hospital bed, its sheets pressed and clean.

How many people had died here, in this room? I gave myself permission to feel the weight of the unknown, permission to be the whole of my unprepared self.

Sharon waited outside the hospital room while I washed my face in the adjacent bathroom. The cool water brought me back into my body, somewhat. I followed her down the hallway, a tunnel of air-conditioning, past nurses and doctors and cancer patients. Sharon punched the elevator button for the main floor. We descended.

"Well, maybe enough story collecting for a day?" She smiled, gentle. "That one really punched you in the gut."

Sharon went to buy a soda from the vending machine. We retreated into an inner hallway toward her office: a windowless, crunched space full of papers. Sharon offered me a seat on the purple couch, and I accepted, grateful to be still.

"Oh, and one more thing," Sharon stated between sips of cola. "Part of listening is paying attention to your emotional reaction. Let yourself be changed by the process. The greatest learning will come when you say: 'I want to be different than I was at the beginning of this journey as a result of seeing it through.'

"Otherwise," Sharon sighed, "you're just playing with sounds, not showing up with your real self. It's a great gamble, this life. We spin ourselves in without knowing what's going to happen."

"Oh! That reminds me," she said. "I want to give you this."

Sharon retrieved an orange plastic dreidel from the inside pocket of her sweater. "With a dreidel, like in life, you have no control. You have to enter into the mystery and take your chances."

I pressed the object into my pocket. Later, on the bicycle, I kept it safe in my handlebar bag.

"Keep on fighting the good fight," I said, thankful.

"No, I'm not a fighter," Sharon countered. "I'm not about words to do with battle. I'm into magic. I'm a big soap bubble."

"A soap bubble? Well, blow good bubbles," I managed, smiling.

"All right," she said.

CONNECTING GEOGRAPHY

Zella Downing, a painter and writer living on the South Island of New Zealand, derives her sense of place from rivers. To her, the river is a connecting geography: the sinew that strings together her past and present. Zella grew up in Colorado. As a teenager, she rode her ten-speed bicycle to a bridge with an inner tube over her shoulder. She parked her bike, jumped off the bridge, floated down the river, came back, got on her bike, and pedaled to work at the A&W Root Beer stand.

Zella's favorite thing to do, other than running up or down the mountains, was floating down the river on an inner tube. "We used to float down the river as a means of entertainment. And there were times the river was kinder to us because we had no sense."

Once, her sister got stuck in a snag—an undertow. Zella's sister had been wearing Levi's jeans because she didn't like to wear shorts; they weren't easy to swim in once they got wet. Fortunately, it all ended well.

Later, Zella moved to Montana and started working at a restaurant. "The river there wasn't as deep and it was much colder and we didn't do as much floating. But sometimes after we closed the restaurant at eleven o'clock at night, after mopping the floor and putting up all the chairs, we would drive up and find this one swimming hole and we'd go swimming. And if we'd met any tourists during the day, we would invite them to come with us. But mostly we would just sit there and talk about all the different tourists during the day. God, that was magic. We'd look at the Big Dipper. It'd be midnight by the time we got home. And then we'd go and do it all the next day," she said.

Years later, she moved to New Zealand. "My connecting geography is the river," Zella said. "I love diving in."

Most of the time, while she swims, Zella thinks about her sister, many rivers away. She uses the river "to help send her happy thoughts. And when I die, I've always thought I'm going to send my ashes to Montana to be spread around the mountains. But I think I'm going to have half of them put in the river," she said. "I think to be able to dive into the river is a blessing that everybody in the world should be able to have."

TOXIC

Perfluorochemicals (PFCs) are a group of chemicals used to make fluoropolymer coatings and products that resist heat, oil, stains, grease, and water.[15] Clothing, furniture, adhesives, food packaging, heat-resistant nonstick cooking surfaces, and the insulation of electrical wire can all contain fluoropolymer coatings. PFCs are also present in flame-retardant foams used for firefighting. Chemicals in this group like perfluorooctanesulfonic acid (PFOS) and perfluorooctanoic acid (PFOA) do not break down in the environment. Instead, they build up in the tissues of wildlife, bioaccumulating to a point of toxicity. Some people call them "forever chemicals" because they do not go away.

PFCs also accumulate in the bodies of humans. In the seacoast area of New Hampshire, Mindi Messmer, a state representative and former hydrogeologist, has been tracking the impact of these chemicals on constituents in Greenland, Rye, and North Hampton since 2014.

At that time, her son Kegan was ten years old. Kegan told his mom about one of his elementary school classmates who had leukemia. There were four other cases of rhabdomyosarcoma, a rare childhood soft-tissue cancer, in the community. Mindi suspected that the cause of the outbreak was environmental. Based on her hydrogeological knowledge and archival research, Mindi concluded that the cancers could be traced back to groundwater that had been contaminated from Coakley Landfill.[16] Coakley Landfill is a superfund site near Pease Air National Guard Base that accepted municipal and industrial waste between 1972 and 1982.[17]

In 1994, a cap was placed on top, but the bottom remains unlined. Chemicals leach into the ground and surface water, impacting the wells of nearby homes. The base used PFCs in the form of flame-retardant foams meant to extinguish fuel fires. These foams were frequently applied in drills.

Local well water samples of three hundred homes surrounding the landfill revealed PFC concentrations at up to forty times the national average. Even low levels of exposure can be hazardous to human health. Mindi helped identify a cancer cluster in the area, mostly of pediatric cancers. The implications of this toxicity extend beyond New Hampshire. Communities in Pennsylvania,[18] Delaware,[19] and Michigan[20] are dealing with similar contamination from PFCs.

Mindi ran for office and was elected to the New Hampshire House of Representatives, where she served from 2016 to 2018. During that time she sponsored three bills, one relative to blood testing for individuals exposed to perfluorinated chemicals in private or public water supplies,[21] the second prohibiting the sale of furniture with flame retardant chemicals,[22] and the third in support of insurance coverage for pediatric autoimmune neuropsychiatric disorders.[23] Mindi also formed an advocacy group called the New Hampshire Safe Water Alliance that pushes for remediation at the Coakley Landfill site and to stop the flow of contaminants into surface water.

The scary part about PFCs is that you have to test for them to even know that they're there. "There's no other way. It doesn't smell," Mindi added. "You can't see it. It's invisible."

WAVES

Sometimes a question is more powerful than an answer. Phil Drawn, a storyteller from Augusta, Georgia, spoke to me about waves. One day, his story began, two waves were out on the ocean, heading for shore at full speed. They had white caps on top. "They are just enjoying life," Phil said.

One wave saw the beach. Ahead of him, the waves crashed against the sand, destroyed. The wave in front said to the wave behind him, "Our lives are about to end on the beach in just a moment."

And the other wave said to him, "Son, let me tell you something: you have been enjoying life so much as part of a wave that you forgot you're just part of the ocean."

In the context of climate change, actions are individual. It's so easy to be caught up in the waves of our own existence. But we are part of a larger ocean, a larger planet. How can we make decisions accordingly?

EVANGELIST

At the end of the 2017 People's Climate March in Washington, DC, I took a moment to rest across from the White House in Lafayette Park. I found a patch of shade. I had documented fifteen stories that day; I was exhausted.

A man sitting nearby noticed my cardboard sign: "tell me a story about climate change." He introduced himself as Lowell Bliss and told me that he comes from a "very conservative evangelical background where environmentalism is a dirty word."

He sucked his tongue against his teeth, then exhaled. "But what would it mean to lop off the 'ism' and get to the root of even just what environment is?"

Lowell first arrived at his definition of "environment" in 2005. He had worked with his wife as a Christian missionary in India and Pakistan for fourteen years. That year, a magnitude 7.6 earthquake hit Pakistan,[24] killing at least eighty thousand people in building collapses and landslides.[25] "Part of the problem was it was all up in the mountains," Lowell told me. "People couldn't get in and people couldn't get out."

At his wife's suggestion, Lowell traveled to Pakistan to join an earthquake relief team at a field hospital clinic set up by the Japanese government. "I'm not a doctor, but I do speak Urdu, so I was able to be the translator," he told me.

One evening the curtain parted at the medical tent. A family carried a rope bed piled high with blankets and quilts.

"My understanding of the environment and environmentalism really was like the lifting of the blanket," Lowell said. "I lifted the blanket and I was looking in the eyes of this fourteen-year-old Pakistani girl, and she was very sallow skinned. She was all kind of bunched up. She was all knobby elbows, knobby knees, and she was just kind of gaping, gasping for breath, like the blankets were water and she was a fish."

The girl had cerebral palsy. When the earthquake happened, she was in bed and a wall of her house collapsed on her. "She had no sense of what happened, why it happened, and it frightened her greatly," Lowell told me. "She didn't have speech, but her father and her brothers and her uncles were all explaining that, you know, she had reverted. She was back on the bottle. She was whimpering all the time."

The doctors examined the girl and told Lowell, as a translator, to say that the earthquake did not cause her problem. "I wasn't going to translate that because they weren't stupid. They knew that," Lowell told me.

Then the doctors told Lowell to tell the family that there was nothing wrong with her. "I couldn't translate that either because it felt like everything was wrong with this situation," he recalled. The doctors prescribed aspirin, encouraged the family to warm her milk and let her be on the bottle for as long as she needed. "Just love on her, really," Lowell said. "There really wasn't anything that we could do." Lowell, being religious, prayed.

"I knew right then and there that that's what environment means to me. Environment is that which is collapsing around the people we love," Lowell said. "The environment is collapsing around the people for whom Jesus died for, the people who Jesus loves dearly."

He noted that earthquakes aren't climate change, yes, but something about his faith opened up in that tent. Five years later, floods came to Pakistan, displacing millions of people. "I remember researching that district during the flood and thirty-five people died. And I always kind of wondered, well, was she one of them?"

Since then, when Lowell encounters people in his evangelical community who are "closed to climate change, closed to environmentalism. I say, 'Listen, you know, Jesus would have us lop off the -ism and get right to the root of the problem.' And it is, you know, innocent people who are suffering because of our actions. So that's my story."

I turned off the tape. "The way you get people thinking of a new narrative is that you help them grieve," Lowell added. "The way you help them grieve is you say, 'Tell me your story.' "

14

CANADA

IGLOOLIK

Igloolik, Nunavut, fourteen hundred miles south of the North Pole, is an umbrella town. The only way to get in or out is by passenger plane, dog sled, snowmobile, or—for a few weeks in summer when the sea ice melts—boat. Around seventeen hundred people live there.[1] The few stop signs in town have words in both English and Inuktitut. People say yes by raising their eyebrows, and no by scrunching their noses.

When I visited in July 2018, with support from a National Geographic Early Career Grant, the sun was eternal: twenty-one and a half hours of it. If I had arrived in June, near the solstice, the sun would never set at all—just circumambulate around us, a bright yellow juggling ball, always above the horizon. In July, there were a few hours of sunset and sunrise all at once. It never got fully dark. I learned to turn off my eyes to fall asleep.

Life in the north is expensive. Fruits and vegetables are flown in; a two-pound bag of grapes can cost more than twenty Canadian dollars.[2] Earlier that summer, there had been a spate of polar bear attacks in communities nearby.[3] People were on edge.

There were also beautiful things. I arrived to the sound of ice melting on the beach, when a few flowers were blooming. There were insects on the hillside by the cemetery. Mosquitoes: one. Spiders: two. Sheryl, my host for the first two weeks, went to collect her water as ice or frozen snow in five-gallon orange paint buckets. She scooped the ice with a saucepan and boiled it back home for consumption.

I visited the Igloolik community radio station, Nipivut Nunatinnii "Our Voice at Home," and had them run an announcement that I was in town and looking for stories. Then, I listened.

WALRUS

Marie Airut, a seventy-one-year-old elder, lives by the water. We spoke in her living room over cups of black tea. "My husband died recently," she told me. But when he was alive, they went hunting together in every season; it was their main source of food.

"I'm not going to tell you what I don't know. I'm going to tell you only the things that I have seen," she said. In the 1970s and '80s, the seal holes would open in late June, an ideal time for hunting baby seals. "But now if I try to go out hunting at the end of June, the holes are very big and the ice is really thin," Marie told me. "The ice is melting too fast. It doesn't melt from the top, it melts from the bottom."

A few years ago, she went seal hunting by boat, and brought the animal onto the land to eat fresh seal meat with her family. The skin looked "really old, and it was very easy to break," she said. She blames this on increasingly warming water temperatures. Caribou hunting has also changed. In the 1970s and '80s, she went caribou hunting on Baffin Island in August. Back then, it was "very, very hot, with lots and lots of mosquitoes. Now it doesn't have any mosquitoes. The water looks colder at the top, but it's melting from the bottom. The sea is getting warmer," she repeated.

When the water is warmer, the animals change their movement. Igloolik has always been known for its walrus hunting. But in recent years, hunters have had trouble reaching them. "I don't think I can reach them anymore, unless you have seventy gallons of gas. They are that far now, because the ice is melting so fast," Marie said. "It used to take us half a day to find walrus in the summer, but now if I go out with my boys, it would probably take us two days to get some walrus meat for the winter." Marie and her family used to make fermented walrus every year, "but this year I told my sons we're not going walrus hunting. They are too far," she said.

"I read my Bible every day, and I know things will change. And I believe both of them are happening now, what is written and what I see with my own eyes."

WARMING

Theo Ikummaq has worked as a wildlife officer in Igloolik since 1982. When Theo was a child, his family was nomadic. In the wintertime they lived in a sod house. In the spring and summer, they followed the animals: caribou, narwhal, walrus. He grew up learning how to hunt and navigate. "I was brought up to care about the environment," he told me.

When it comes to climate change, he said, "The big thing that nobody is really aware of is the temperature change of the water. That's what is creating climate change. Not the sky. Not the land. Water," he said. Theo pointed out to the bay and told me that the ocean floor, fifteen to twenty years ago, was averaging –2 degrees or –2.5 degrees Celsius. Salt water doesn't freeze until roughly –2. "Today, any time during the year, it's above zero," he said. "Everything at the ocean floor is thawing."

While people in town might not notice these changes, the hunters do. New birds come to Nunavut annually, and the diversity of sea creatures is shifting, too. "Seals are scarce," Theo said, which tells us that "the food source of the seal is somewhat diminished." Humans, polar bears, foxes, and wolves all rely on the ringed seal for food.

"Whatever happens in the sea affects the land. Whatever happens on the land affects the sea," Theo said. "If you look after the whole system, the whole system looks after you. That was the theory the Inuit had at one point. We're somewhat removed from that because we had to become like the rest of the world, to a certain degree. Other cultures coming in affected our culture. The culture coming in was stronger. We had to follow it. It was forced upon us, more times than not."

Theo described climate change by saying, "the world shifted." He started to notice this shift in the early 2000s. One example is the wind. When he was a child, the northwest wind was predominant, and it created a pattern of distinctive ridges that people could follow in navigation. Hunters would leave camp and follow patterns created by the wind in the snow. Later, when the wind had erased their tracks, they could return to camp by following the pattern of the ridges in reverse.

But now, the winds are less predictable. Starting about fifteen years ago, "When our elders were navigating by snowdrifts only, they were getting dislocated. They ended up at the wrong place. They weren't lost. They just ended up at the wrong place and then corrected their bearing," he said. "The youngsters, with their GPS, were getting to the place where they had to go."

Sightings of killer whales have increased throughout the territory in recent years.[4] "Because the ringed seals have never seen a killer whale before, they don't look at it as a predator, the ultimate predator," he said. "They're not even afraid of it." As a result, killer whales go from bay to bay, wiping out everything. "It's one killing machine that's coming into our neighborhood," he added. "It's not just the humans; the animals aren't aware of what's happening out there."

JOURNEY TO GREENLAND

Michael Immaroitok is a regional lands administrator for the Government of Nunavut, a position he has held since 1994. As a boy, he lived with his family two hundred miles from Igloolik, in the "middle of nowhere on Baffin Island," he said. Although they were in a remote area with only four other families nearby, accessing food wasn't a problem. "There was a whole lot of caribou on Baffin Island back then. We had no problem catching caribou and there were lots of sea mammals. Lots of fish," he said.

Michael's parents taught him to hunt, skin animals, prepare the skins, preserve the meat, and sew. Though he now has a respiratory problem that prevents him from hunting as much as he would like, he still fishes for Arctic char in the summer with his brother. He knows the area well but uses a GPS as a backup.

In 1987, Michael joined a three-month dogsled trip from Igloolik to Greenland. While they were traveling over a glacier, some of the dogs fell into a crevasse. They were able to rescue them using ropes.

"Each area that we traveled on had a different kind of snow. Snow forms differently in certain areas," he said. "We were retracing the route of our ancestor, who was running away from a shaman." In one place along the way, that ancestor starved. Michael and his team wanted to have tea there, but when they lit the Coleman stove, the dogs started barking at the air. They were too nervous to sit still. When Michael moved just two hundred yards away, the dogs calmed down, and he was able to enjoy his tea.

"It was weird," he said. "There's some spirits up there." The spirits, he added, are neither bad nor good. They are just there.

POLAR BEARS

Francis Piugattuk has worked for twenty years as a wildlife technician at the Igloolik Research Center, a government-owned building on top of a hill that resembles a giant, white mushroom. It was built in the early 1970s as a place to bring together Inuit knowledge and Western science. As a wildlife technician, Francis processes samples of polar bear bones and tissues and produces research permits. In the lab, he analyzes fat samples, ear tags, and tattoos to help track polar bear hunting throughout the territory. Polar bear teeth, Francis told me, "have growth rings like trees," Suctioning out a tooth and counting the lines helps age the bears.

When Francis was a child, polar bear sightings were infrequent. "Even seeing tracks was an anomaly, a cause for excitement. And if people wanted to harvest polar bears, they would have to go long, long distances," he said. Up until twenty years ago, the only animals attracted to walrus meat caches were Arctic foxes. Now, the community is setting up electric fences and trying to extract the fermenting meat before the polar bears can get to it. While the population of polar bears hasn't technically increased, they are moving closer to human settlements as ice patterns change. About 16,000 of the 20,000 to 25,000 bears in the world's polar regions live in Canada.[5]

Francis acknowledges that Western science and traditional Inuit knowledge are two systems that "seem to be at odds continuously." When he was young, his parents waited until the last day of school in June to bring him from Igloolik onto the land for the summer. Then they followed the hunt until school started again in September. His elders would pass on lessons about which water was safe to drink—free-flowing was better than still. "Even though they did not learn what they knew in school, like we did, they learned. They had years of existence to learn," Francis said.

Elders, Francis told me, were able to live sustainably off the land by selling fox or seal pelts in exchange for rifles, boats, and other materials. Today, it's only those in the wage economy who can afford to buy an outboard motor or ammunition. "The cost of living is so great now that it's not even viable to try to exist as a hunter," he explained. "Those of us that do not hunt live on pasta and macaroni, rice, soup: food that is not as nutritious. Those that are still able to afford it are now going out and acquiring country food." Country food, I had learned, includes traditional food such as bannock, arctic char, eggs, and muktaaq; it is often shared as a gift between families and in the community.

Climate change, Francis told me, is already here. "The ice used to stay longer," he said.

COUNTRY FOOD

Terry Uyarak, a hunter in his early thirties, has deep tan lines around his eyes in the shape of his sunglasses—the sign of a summer spent out on the land. He invited me into his kitchen, where we ate *maktaaq*, frozen pieces of whale skin and blubber, and *tuktu*, caribou meat. Terry's wife, Tanya, cut the meat with an *ulu*, a knife with a semicircular blade and a handle that is the sole provenance of women.[6] I liked the rhythm of its rocking, the rounded edges.

Every season brings something new: beluga, narwhal, caribou, arctic char, walrus. Terry works for the Government of Nunavut, coordinating programs that teach hunting to youth, and document elders' hunting methods. He is a leader in his community. "Usually in early summer, there's no wind," he said, noting that hunting is easier when the water is calm and there is less ice. The high winds that day had prevented him from going out fishing. He also noted that, when he was younger, the ocean would freeze in late September. Now, come Halloween, he can still go boating. In the past, in late October, he would be driving a snowmobile.

"It's changing quite rapidly. And I'm not old at all. I'm thirty-one, and I can tell very much how it changed," he said. Terry told me that polar bears are also coming closer than they used to, a threat to stored food. "Now we have to be armed all the time on our camping trips," he said. He tries to be careful, even in the winter, to observe the ice and make sure that it is not too thin.

When hunting is less reliable, his family has to buy more groceries from the stores. "It's very expensive—very, very, very, expensive—for us here."

Later, I rode on the back of Terry's Honda ATV to the place outside town where he keeps his dog team. We tossed them pieces of raw fish: Arctic char, the leftovers from his most recent catch. Terry's face looked more complete with his sunglasses on. As we watched the dogs eat, I thought about the delicious caribou meat we had shared, still fresh on my tongue. Terry had warned me that I would crave the meat later, and he was right. The animal ran through me. All I wanted was more.

Country food is very nutritious, and also expensive to harvest. Consider ten thousand Canadian dollars for an outboard engine, then add a boat, snowmobile, bullets, gun, the cost of gasoline shipped in from the south, an ATV. Many people can no longer afford their traditional lifestyle. Sharing the bounty is the norm and a necessity. Once the meat is distributed, it is time to harvest more.

INFORMAL WATER

Sonia Wesche is an associate professor of environmental studies and Indigenous studies at the University of Ottawa who has worked collaboratively with Indigenous communities in Yukon, the Northwest Territories, and Nunavut for more than a decade. Her goal is to help these communities adapt to environmental change. That means taking a close look at food and water security. In summer 2018, Sonia visited Igloolik and Sanirajak with her colleague, a freshwater ecologist. At the time, Sonia was forty-two years old. Compared with other communities she has worked with in the Arctic, she found Igloolik both more remote (not accessible by road) and more connected to traditional language.

"One of the things that really struck me in visiting these two communities was how important the informal, or unregulated, water system is. And this was very different from the way we use water in the more urban, connected south of the country," she said. In Igloolik and Sanirajak, people covet access to water that is not treated through a municipal system. Elders prefer water that comes from the land as snow or ice. When they steeped a bag of Red Rose Tea, the tap water would turn it dark brown, but tea made from untreated ice or snow was a soothing red.

People go to great lengths, in time and money, to access that water. Not just any snow or ice will do—you have to know where to go. This is remarkable in a place where resources are already slim, and many families are food insecure.

Climate change has made it more difficult for communities to maintain municipal water systems. Unlike Canada's south, the ground in Nunavut is frozen. Permafrost means that communities cannot rely on groundwater systems; surface water is the only source. Climate change has translated to increased evaporation and shrinking areas of surface water. Ponds dry out. Some sources of ice—especially floating, multiyear ice that has lost its salt content—are less common.

As Arctic communities grow, Sonia said, it's crucial to consider how population size will impact water resources. The important thing, she added, is "to ensure that the people in the north have quality water for the future."

GLACIER WATER

As July progressed into August, the ice in the bay broke up into pieces that floated in and out with the tide. Some pieces of ice would stay onshore, melting monoliths, taller than me. Others would be retrieved, hours later, back into the water's warming embrace.

Marcus Angutiqjuaq, a father in his early twenties, used to go hunting frequently when his grandparents were still alive. They traveled four to five hours by boat in the summers to Baffin Island. Their campsite was next to a glacier with a river flowing down. When he was eight years old, Marcus remembers watching an elder go back and forth to the glacier many times a day to fetch water.

"It's completely different from tap water. It's very smooth. I'd rate it best of all the waters," Marcus said. In town, Marcus and his family get water from the tap. Sometimes, when they have extra money, they rent an ATV to drive to a part of Igloolik where it's safe to fetch icy snow, scoop it into a plastic barrel, and carry the ice back home to the kitchen, where it melts into water.

Marcus lamented the proliferation of trash on the shore. "I used to play out near the seashore, and there wasn't any garbage. But today there's a lot," he said. "People throw things on the ground instead of keeping it until there's a garbage can nearby."

INFRASTRUCTURE

Joasai Kublu is the director of public works for the municipality of Igloolik. We spoke in his office, under fluorescent lights. A map of all the homes on the island was tacked to the wall.

"Our reservoir was very small before. We ran out over the winter. By spring there was no water in the lake and we had an emergency," Joasai told me. Two years ago, the system was upgraded. A larger reservoir was dug. A cargo plane arrived with flexible pipes and pumps for the installation. Contractors arrived from Quebec to install a new filtration system.

A pump on the bottom of the reservoir now moves water to the pump station, where it is treated. The pump station receives electricity from the power station, which is fueled by diesel motors. A ship arrives once a year, in early September, to resupply the fuel tanks.

After filtration, the reservoir water is moved to the water trucks, which drive to different parts of town, filling tanks. Each home has a water tank that holds around 350 gallons of water from delivery. Larger buildings, like the hamlet office, can have tanks as large as 4,000 gallons.

SEAL

I spoke with Leah Angutiqjuaq, age forty-two, in her relative's home in Igloolik. We had just boiled water for tea. The most pronounced climate impact, for Leah, is in the timing of the seal hunt. "The weather is changing," she told me. "We used to go out seal hunting for one to two months. It's only three weeks now." When she was younger, her family camped and spent time on the land. "Now it's different, because we need money and we hardly have any dogs. We have some, but only as pets now."

"Our older people have passed away," Leah added. "We can only buy food now. We used to share. If we went out camping, the family would come. Now it's different." Without a dog team, hunting is prohibitively expensive. "They try and let young ones go out camping, but they need money," Leah said. "Many years ago, they used to help each other without money." In a town where many people make minimum wage, a subsistence lifestyle is often out of reach. "Too much money now, maybe," she said.

Before I left town, Leah sold me a white ring carved from a walrus tusk. The carving was in the shape of an owl, its wings spread wide around my finger. After I paid her, Leah went straight to the grocery store, cash in hand, to buy food.

HUNTING DOGS

Maggie Amarualik is an Inuit woman who moved alone from Igloolik to Alberta when she was in her late teens. In 1959, a few years before Maggie was born, the Royal Canadian Mounted Police tied up, shot, and killed her father's entire team of hunting dogs, effectively ending his life as a hunter. "I remember my dad crying a little bit as he talked about how much he loved these dogs. They listened to every command," Maggie said.

In her 2016 book *The Right to Be Cold*,[7] Inuit environmental and human rights advocate Sheila Watt-Cloutier describes how *qimmit*, or sled dogs, were shot. "While the official explanation given at the time was that they were culled to prevent the spread of distemper and attacks by sick dogs," she wrote, "many now suspect that the destruction of the dog teams was another way to force Inuit families to move from outpost camps into settlements, by removing their only mode of transportation."[8]

An additional source of community pain is the history of Inuit children being taken from their families and educated at residential schools in the South. At twelve years old, Sheila Watt-Cloutier was sent to Churchill Vocational Centre in Manitoba, a residential school for students age twelve to seventeen. The dormitories had previously been army barracks. "The program taught academics as well as vocational skills like cooking, child-rearing, and sewing for girls, and trades such as carpentry, mechanics, and welding for the boys," she wrote. "And while the skills (especially those taught to the girls) were similar to the types of things Inuit children would have learned with their families and in their communities, the instruction was in southern ways," she added. "We were being deprogrammed from our Inuit culture and reprogrammed for the southern world."

Residential schools, run by the federal government, existed in Canada for nearly a hundred years; the last one closed in 1996. Approximately 150,000 Indigenous children were taken from their families to be reeducated in English or French. "Resistance was rewarded with punishment, and many students experienced physical, emotional, and sexual abuse," Sheila wrote. "Today, people are still trying to heal from these horrific experiences." When children were abruptly removed from their communities, families were unable to pass on tradition and culture. "This has resulted in generations of trauma," she explained.

Older people in the community were told that if they came to the hamlet and started working, they would have everything that they needed: free housing, good homes, and a place to stay warm. "They said they were going to give everything to the Inuit in order to survive," Maggie told me. "But really look at what we have now. We're starving more, and not being heard or listened to. So we're starving for attention," she said, laughing in the way that's only audible when something really, really hurts.

"We became very deathly afraid of the white people, because they could shoot things that were really important to us, and then they took our brothers and sisters away from us," Maggie said. "We couldn't hunt anymore."

The intergenerational pain in Igloolik is unusual, perhaps, in how recent it is. Elsewhere, it's buried under more layers of time. In Nunavut, these wounds are still fresh.

NAVIGATION

Serapio Ittusardjuat lives in a house with his children and grandchildren, near the hockey arena. When I asked him his age, he laughed. "I'm seventy-three and still young." Serapio was born in Igloolik and raised as a hunter—some skills he learned from his father and grandfather. Others he picked up from careful observation. "We know the movement of the walrus by season," he told me. "They only eat when there's a tide."

In the Arctic, navigation is a matter of survival. The winds used to be more consistent, which would let a hunter know where to go based on the direction of ripples on the snow. Serapio knows the currents around Baffin Island and which cracks in the ice to avoid. And, of course, he never used to navigate using a GPS or a snowmobile. "It's not safe to travel alone on the machine. Broke down, what can you do?" he said. A dog team and a story were all he needed. "We used to be told to go to that place by telling us where to travel and we got there," Serapio told me. Place names were important—they told a hunter what to look out for.

The best way to teach kids, Serapio told me, is to demonstrate these techniques firsthand. "I used to do a little bit of work, and my children would be watching, watching me, and I just tell them what to do. They learned fast," he said. And after a certain point, the teacher lets go. A kid has to try and fail on their own. "You learn by yourself," he said. "When you can do it, it's all yours."

FOOD WEB

When we met in Igloolik in 2018, Marie-Andrée Giroux was an assistant professor of environmental sciences at Université de Moncton in New Brunswick. She first visited the island in 2011 and lived in Igloolik for two years continuously. Since then, she has returned to the Arctic for a few months each summer to conduct research. Climate change is more pronounced in the poles than it is in lower latitudes. For Marie-Andrée, melting sea ice is the most pressing concern in the circumpolar region related to climate change. In the north, sea ice isn't just a natural element—it's also an infrastructure used for traveling to hunting grounds. "When the sea ice melts earlier or the conditions are not as stable as usual, it's like the roads being unstable and unpredictable. So it has a big influence on traditions," she said.

Wildlife has the same problem. Many species, like Arctic foxes, cross between islands and the mainland using sea ice. In the winter, an Arctic fox can travel for thousands of kilometers across the region, often following polar bears who prey on seals. After a polar bear leaves the seal carcass behind, a fox will scavenge and eat what remains. If sea ice conditions are unpredictable or melt earlier than usual, a fox's access to food and water—and its ability to reproduce—is limited.

It's easy to think that sea ice would impact only the ocean, but there are many energy exchanges between the terrestrial and marine ecosystems. Seabirds, for example, nest on an island, forage in the water, and then come back on the land, where their guano fertilizes plants. The tundra, as a low-productivity area, relies on energy inputs from the marine environment. This means that when sea ice dynamics change, not only marine food resources but also terrestrial resources change. And because people depend on terrestrial resources, whether by picking eggs or eating caribou, what happens to the sea ice impacts the human population, too. Everything is interconnected.

Still, the specifics of climate impacts on this system are difficult to predict without further study. "Right now it's pretty hard to predict based on all those intricate relationships which are just being described right now," she said.

One key species that is being affected by climate change in the tundra is the lemming. Lemmings are small rodents that live, during the winter, under the snowpack, where it's warm enough for them to survive and reproduce. The snowpack, in addition to insulating their food, also protects them from predators.

Climate change wreaks havoc on this delicate balance. When the melting and freezing cycles change, the snowpack that lemmings rely on becomes less predictable. In a rain-on-snow event, the water percolates through the snow and freezes the vegetation underneath, rendering the lemmings' food supply inaccessible. Many predators in the Arctic eat or select their breeding ground based on lemming abundance, and those same predators also eat birds and bird eggs. On Igloolik, when there are more lemmings, Marie-Andrée has observed that Arctic foxes and avian predators (such as the long-tailed jaeger, parasitic jaegers, gulls, ravens, snowy owls, and other raptor species) are more abundant. When climate change impacts the lemming, it indirectly impacts other species in ways that are not yet fully understood.

Marie-Andrée is most energized by climate solutions that take into account the needs and interests of different groups involved. Snow geese, which migrate to the Arctic from the United States and Canada to breed, have increased exponentially in the last four decades due to an increase in the amount of agricultural land where they feed during the winter and along their migratory path. "They have increased to a level where they are detrimental to Arctic ecosystems. When they come here to reproduce, they overbrowse the vegetation," Marie-Andrée said. This destroys the habitat, and forces predators to eat other birds at higher levels.

One approach to this problem is to implement snow goose harvesting programs—not only though a spring hunt in the south, but also by encouraging egg collection and harvesting of adults in the north at their breeding ground.

"If we can work toward supporting harvesting programs which are beneficial for conservation issues at the same time, I think that's really good," she said.

SASQUATCH

The vast majority of Canada's population, two out of three people, live within a hundred kilometers of the US border.[9] In Nunavut, a territory with a population of just under forty thousand people, anyone who lives south of the Arctic circle is considered a "southerner."[10]

I met one of these southerners, Hunter McClain, on the street in Montreal. Hunter is from a small town in northern British Columbia, close to the Hudson Bay Glacier. The glacier, which used to be visible on the mountain, has been receding to the point where it's nearly invisible in summer and spring. "People who live out in the country are pretty in tune with the seasons, and we noticed changes in the wildlife," she told me. "The wildlife has been going a bit nuts."

One year, the bears didn't hibernate because they couldn't find enough food. "All the juvenile bears over the winter were running around town looking for food. You could see them losing hair and they looked so thin," Hunter said. "I had never seen a really skinny bear before, but when you see a skinny bear loping around and standing up, you really realize that that's Sasquatch." The bears on their hind legs looked like the legendary monster. Hunter was terrified, and equally "weirded out by people who live in that area who are climate change deniers." To her, the connection to climate change was indisputable.

15

PERU

THE LIVING LIBRARY

"Walking through Farmacia is like swimming in plants," Sophia Rokhlin said. She was trying to reheat lentils in the kitchen of a shared house called Ametra in Pucallpa, Peru, but the stove was running out of propane. Rich, humid Amazonian air flooded the kitchen through a slatted window. The delicate spines of spider webs glittered in the sun. An orange cat padded across the floor. Everyone moved slowly in the heat. Sophia stirred the lentils with a wooden spatula and eyed the phone number for the gas delivery guy, tacked to the wall.

Sophia was one of ten volunteers in the house working for Alianza Arkana, a nonprofit organization that conducts regional environmental, health, and education projects, among them Farmacia Viva Indígena, or the Living Indigenous Pharmacy.

Farmacia is five hectares of Amazon rainforest (roughly the equivalent of nine football fields), preserved as an intact library of indigenous plants, many of them medicinally useful. In 2017, the residents of nearby river village Paoyhan invited Alianza in to help with the project, making Paoyhan the center of a small, international experiment. The goal is to address challenges in both the community and the larger region: climate change, the disappearance of native cultures and languages, and deforestation that threatens biodiversity.

Traveling from Sophia's home in Pucallpa to Paoyhan involves a boat trip. There are two options: the fast boat—70 soles (US$20) for three hours; or the slow boat—30 soles (US$9) for five hours. Both boats are loud: a vibrating hull with the motor as its node, *cumbia* playing at max volume. They are invariably filled with passengers or cargo, often bananas. Sophia and her housemates make this trip regularly.

On a recent visit, Sophia told me, the group trekked through Farmacia with their raincoats on, hoods zipped up, even under a clear sky. "Just our eyes out," she said. "Yes, it was hot, but worth it not to be eaten alive." The mosquitos, Sophia said, "bother new people the worst. After you've been there for a while, it's not so bad." She turned back to the sink, found a spoon in the drying rack, and started eating her lentils cold, straight from the saucepan. "The guides were telling us the names of plants faster than we could write them down," she told me. "This is why the list is a bit disorganized. But it's getting better."

In January 2018, the entire Ucayali River basin was inundated by more than mosquitos. Seasonal rains had come a month and a half late, and with them the floods. Boats still departed from Pucallpa, but the village of Paoyhan wasn't receiving visitors. Its houses, wooden and mostly unpainted, are roofed with palm fronds and built a meter or more off the ground to stay dry when the river grows. During a flood, people travel from house to house via boat, or by wading through thigh-high water. In the rainy season, drinking water becomes a concern.

People in Paoyhan pump and collect groundwater for drinking. In a flood, raw sewage from dry toilets mixes with the water. Stomach infections are common, and recourse to the clinics in Pucallpa is limited. In this region, as in many around the world, climate change, water, and public health are solidly intertwined.

The weather has become fiercer and more erratic in recent years, exacerbating these difficulties. In the past decade, the Peruvian Amazon has experienced the most intense flooding and droughts in recent history.

Climate change is likely to increase the frequency and severity of these floods. Changes in rainfall patterns in the Peruvian Amazon are expected to accelerate soil erosion, reduce water availability, lower crop yield, and increase disease. It is exactly these struggles that Farmacia hopes to address.

MEDICINAL PLANTS

The region of Ucayali is home to five hundred thousand people, a constellation of settlements that hug the banks of the Ucayali River, which flows through the forest on its way to the Amazon River. The majority of the floodplains in Ucayali are, for now, still forested. The Ucayali River, broad and brown from suspended sediment, is a highway not only for people and bananas, but also for the cut trunks of trees—some cut legally, many poached—lashed together into rafts and floated to sawmills downstream.[1]

Pucallpa is the regional metropolis, but smaller towns like Paoyhan and hamlets like Santa Elisa, population three hundred, are spread throughout the forest. Many of the smaller communities resemble Santa Elisa, a long hallway of single-room, wooden houses bookended by banana plantations. They are lit for a few hours each night by a string of streetlights powered by a generator, every third one out. By day, shade is a luxury. Along the riverbanks, under broad-leafed trees, there are one-plank benches and places to hang hammocks. Life here is gendered—men farm, fish, and hunt; women raise the children and embroider fabric under the trees. A few of the homes have TVs.

Some people in Paoyhan make a living by selling plants like cat's claw (*uña de gato*), a medicinal vine used to combat viral infections, cancer, arthritis, parasites, and inflammation. A large fraction of income comes from women's artisanal work. Women of all ages often spend their days, between cooking and taking care of the house and kids, working on embroidered maps of rivers—geometric patterns that tell a story of the land and water and the people who live here.

The culture of these villages is Shipibo-Conibo, an Indigenous group in the Peruvian Amazon who live between 6 and 10 degrees of latitude south of the equator. Shipibo people have been living in the Ucayali region of the Amazon for thousands of years. They speak Shipibo, a language in the Panoan language family, though most people under seventy also speak Spanish. Shipibo culture is distinctive for its connection with traditional plant medicine. Shamans apprentice themselves to a plant in order to heal members of their community and themselves. Cut down the trees, and the Shipibo people are threatened, too. To kill the plant world is to kill the Shipibo spirit world as well.

When children and babies in his community have bronchitis, pneumonia, or vomiting, Victor Sanchez Valles told me, "we cure this with traditional medicine. Medicinal plants. If you can't with those, you have to go to the medical post" in La Libertad, twenty-five minutes away from Santa Elisa by *motocarro*, a motorized tricycle with a bench for passengers or a flatbed in back. Victor, fifty-six years old, is the lieutenant governor of Santa Elisa. He has lived in the town, an entirely Shipibo community, all his life. Victor said that his ancestors and the elders in the community know about medicinal plants, but many of the younger people don't.

Victor credits this disconnect to the current education system. "Because they study at school, primary school and nursery school, some know but not many. They do not know about medicinal plants. And things have changed. They change their understanding with the influence of other countries." In the 1950s, a Spanish-language education system was introduced in Ucayali, creating a gulf between youth, who were educated in Spanish, and the elders in their community, who spoke Shipibo and had extensive knowledge of the plant world. Valles learned about medicinal plants from his grandparents.

In Santa Elisa, medicinal plants don't grow among the banana plantations or in town. "There are none here," Victor said. "You have to go look for them in the center of the forest." Each way takes forty-five minutes to an hour on foot, with a machete in hand. Victor brings the barks of the plants he knows back to Santa Elisa and boils them down to create a drinkable tonic.

When Victor goes walking in the forest, the haul can be unpredictable. "Sometimes there are no plants. Sometimes there are. That's how it is," he said. Lack of access to medicine, both traditional and Western, is part of what inspired the residents of Paoyhan to found Farmacia.

Farmacia is a short distance by water from the community of Paoyhan. To visit the library, people take a long, narrow boat that can carry four to fifteen people. From September to December of 2018, approximately fifty tourists visited. The boats are constructed out of tree trunks, seat two people across, and have a long arm of an extended outboard motor, called a *peke peke*. The trip from Pucallpa to Farmacia is about twenty minutes long, and winds through a small river, so narrow in places that you can reach out and touch the plants on the shore.

In 2017, Humberto Rojas, the elected leader, or *jefe*, of Paoyhan wanted to confront a problem. He noticed a trio of bad things affecting his community. The weather was getting worse. Unpredictable flooding washed away crops, punctuated by periods of drought. Economic opportunities, beyond subsistence farming and crafts, were slim. Some people in Paoyhan had turned to logging to earn a living, which took them away from their community for extended periods of time. While logging brought in some money, it also destroyed the forest elsewhere. Finally, and perhaps most concerningly, Victor noted a disconnect between the youth in his community and their traditional medicinal culture—*non rao*—of healing with plants. In Shipibo, *rao* means "medicine." *Non rao*, "our medicine," refers to plant medicines. *Nawan rao*, or "foreigner's medicines," refers to pills and other forms of Western medication. Without youth to carry the tradition forward, the cultural knowledge of healing with medicinal plants was at risk of being lost.

In early 2017, Humberto approached Alianza Arkana in Pucallpa to ask for help. Alianza had done several workshops in Paoyhan in the eight years since its founding—most recently a women's empowerment workshop focused on stopping violence against women in the community. This included a leadership workshop for young women, and a workshop for young men about sexual health, masculinity, and personal development.

The longstanding relationship between the town of Pucallpa and Alianza Arkana has centered on its plant life. In 2010, a year before he cofounded Alianza Arkana, Dr. Paul Roberts, a professor at the University of Guadalajara and head of the Department of Leadership Studies at the National Institute for Public Health, went to Papa Gilberto, a Shipibo healer in Paoyhan, to "diet" the local plants. Dieting is a ritual in which a person consumes part of a medicinal plant and either fasts or eats only foods without salt, oil, fat, or meat for a certain period of time, to strengthen the effect of the plant's chemicals. Ayahuasca, a psychoactive brew known to activate repressed memories, is one example, but there are many others.

Elías Medina, now treasurer of the Farmacia Viva Indígena committee, describes dieting a plant as a form of communication. We sat across from each other under the screened-in porch behind the Ametra kitchen in Pucallpa. He rocked on a green mesh hammock; I balanced on a narrow bench with one plank missing, and a pillow made to resemble the cross-section of a tree.

"There are plants that you can bathe with, or drink, or diet," he told me. "When you drink, you diet, and you see them"—the plant guardians or owners—"in visions."

Elías paused to put both of his feet on the wooden porch floor, turning toward me. "Medicinal plants are the same as human beings. If you want to delve deeper into the project, you must see it."

This communication happens without talking. "It is like a dizzy feeling," Elías said. "You sleep and you dream. Through your dreams you can see." The guardian spirit of the plant can take the form of male or female. The plant will ask questions in your dreams, he said, questions like, "'Why have you drunk me? Why do you need me?' and in your vision you have to say . . . 'I need you to heal me, and I need your power, I need you to help me.'"

Elías recounted a time when he had rheumatism. He dieted a plant, *machinga* (*Brosimum utile*), a large tree with serpentine roots whose white sap can be used to heal different forms of physical trauma. "In my dreams I had surgery," he said. "The next day I felt my body was good. In three days I did not feel any more pain."

Another plant—known in Shipibo as *janin*, in Spanish as *Tangerana*, and in Linnean classification as *Triplaris americana*, of the Polygonaceae (buckwheat) family, has multiple uses. In Shipibo communities, the leaves can be used to alleviate sunburns, the trunk cortex for malaria, and an infusion of the cortex for diarrhea.

Boaenf, in Shipibo, *ajo sacha* in Spanish, or *Mansoa alliacea* in Linnean terms, of the Bignoniaceae family, smells of garlic and looks like a vine with purple flowers. Its cortex, leaves, stem, and roots can be used to make a pain-relieving tonic that's also useful against arthritis, headaches, epilepsy, fevers, and rheumatism. Filling the house with smoke of its leaves will drive away bats and insects.

Through Alianza Arkana, some researchers had come to do interviews in Paoyhan about medicinal plants and climate change. In 2017, the *jefe*, Humberto, came to Alianza Arkana's headquarters in Pucallpa to discuss the connections between climate change and medicinal plants.

In response to Humberto's suggestion, Laura Dev, a PhD student at the University of California, Berkeley, who also serves as Alianza Arkana's research coordinator, helped plan a climate change workshop in Paoyhan. It was there that the idea for a living library of medicinal plants was born.

The workshop was held inside the local community hall, an aquamarine-blue building filled with natural light. It rests on an elevated platform that protects it from the Ucayali River water. A mixed group of teenagers, adults, and elders came together over a lunch of rice, beans, and fried plantains to discuss the issues.

They drew a map to organize their thoughts. Community members called out issues that they had, and two teenagers plotted each topic with colored markers to a specific place on the map. "A lot of concerns were environmental," Laura remembered. "The bank is eroding, there's flooding, and the lakes are drying up." Another cluster of concerns were about medicine. "We need more medicines for the health post." Or technology: "more technological education in our schools."

Laura consolidated this list based on where they were drawn on the map. Then they ranked each issue in order of importance: low, medium, or high. People who live in Paoyhan presented the problems, and then used the place-based framing of issues to brainstorm solutions.

The key idea that emerged was a botanical preserve—a place where existing indigenous plants could be cataloged, protected for future generations, used to educate youth in the community, and eventually attract ecotourism from the outside.

During the meeting, Segundo Franco, an elder in Paoyhan, jumped to his feet and said, "If I see a botanical preserve in this community by the time I die, I can die happy. I'll just die right there."

The community of Paoyhan chose to designate five hectares of primary forest that was close enough to their community that people could access it via a twenty-minute boat ride. The land had never, to anyone's knowledge, been cut, save for a logging road to elsewhere that bisects it. To manage the land, the group appointed a Farmacia Viva Indígena committee, composed of ten community leaders from Paoyhan. Segundo is its president.

When I visited in 2018, the committee was still figuring out the best way to defend the land from future land-use changes. There were two ideas: to persuade the regional government to grant legal permission, or to ask Servicio Nacional de Áreas Naturales Protegidas por el Estado, a federal agency, to declare Farmacia a protected natural area.

To date, Alianza Arkana and the residents of Paoyhan have documented over four hundred medicinal plants in 3.5 hectares of Farmacia. On each trip to Paoyhan, Alianza Arkana helps to archive more. They divided the land into subparcels and traveled with a Shipibo guide, taking notes to catalog each medicinal plant in Farmacia, like books in a library.

This information is later imported into a spreadsheet. One spreadsheet column lists the name in Shipibo (*bonxix*), in Spanish (*copaiba*), and in scientific terminology (*Copaifera officinalis*—of the legume family). The last column lists uses of the plant. The essential oil of bonxix is a laxative, diuretic, and stimulant. It can be used for the treatment of inflammation, hemorrhoids, cystitis, chronic diarrhea, and against colds and bronchitis. It's irritating if taken in large doses. Another example, called *moe* (Shipibo), *ishanga* or *ortiga Brava* (Spanish), and *Laportea aestuans* (Linnaean classification), is of the family *Urticaceae*. An infusion of its leaves is a diuretic and laxative. The juice of the leaves can be used against conjunctivitis and as a bactericide and anti-inflammatory medicine.

Farmacia Viva Indígena is a cultural survival initiative as much as it is a health initiative. "The Living Pharmacy emerges from us, from valuing our traditional plants. Before, our grandparents did not use chemical medicines," Elías Medina, treasurer of Farmacia, told me. "We used natural plants. Today, this important wealth is lost in our Shipibo culture, which is why the Living Pharmacy is born."

Elías's goal is to have knowledge of traditional plants be a part of the curriculum for kids at school in Paoyhan. "In nursery school, primary school, and secondary school, we want to take them to the Living Pharmacy and explain to them each medicinal plant. What is its use? There are plants for being healers, good doctors, massage therapists, and more. And sometimes our children do not know traditional stories either. This is why we have done this project. So they do not lose the culture," he said.

Nearly one in four Western pharmaceuticals come from rainforest plants. Vincristine and vinblastine, two drugs that originated from the Madagascar periwinkle (*Catharanthus roseus*), are widely used to treat lymphoma and Hodgkin's disease. Taxol, which is derived from the bark of the North American yew tree, is a common treatment for ovarian and breast cancer. The National Cancer Institute estimates that 70 percent of anticancer plants identified so far come from the rainforest. Some doctors like Drauzio Varella, an oncologist in Brazil who has dedicated his life to finding cancer cures after his younger brother died from lung cancer in 1991, scour the Amazonian rainforest looking for future cancer treatments. Since 1995, Drauzio's team has gathered more than two thousand extracts from plants and trees in the rainforest.[2] In partnership with São Paulo's Sírio-Libanês Hospital, Drauzio and his team dry the plant samples and grind them into a powder for testing on tumor cells.

The hope of attracting research projects like Drauzio's to Farmacia offers peril along with promise. While the Shipibo desire to make knowledge of medicinal plants public, they also want to keep this knowledge in the community, safe from large political and corporate interests. As many traditional communities have already learned, it's not an easy trick to pull off.

DEFORESTATION

Plants in the Amazon are especially vulnerable to changes in precipitation and temperature. In Paoyhan, this fact is personal. Shipibo residents feel the climate around them is in flux, threatening their medicinal plants.

"We have felt it," Elías told me. "Heat. The rivers, the waters dry, the lakes. There is fish scarcity and in the middle of December, sun. Before, there was rain. We say there is a total change in our world," he said. "And we know why. Because we deforest."

Climate change is a global problem with highly localized impacts. The residents of Paoyhan have taken a local approach to confronting the changes in temperature and precipitation that they are already experiencing at home. Another goal of the project is to enhance health and quality of life for people in the community by revitalizing and seeding younger community members' connection to their traditional culture of healing illnesses with plants found locally in the forest. Farmacia committee members recognize that planetary health and public health are interconnected. To take steps to mitigate climate change is to enhance the health of people in Paoyhan in the same breath.

"This project is about saving our forest," Elías said. "Not taking down big trees because of global warming, from which many diseases can come. For our children that come after us, we want to explain why it was born, why is this pharmacy important. So they value it and become leaders that can orient their children. Within the perimeter of the pharmacy, nobody will take down big trees, because they are the lungs of the planet."

Elías's concept isn't fanciful. Evapotranspiration, the movement of moisture from forest-shaded soil and through breathing foliage, is a key driver of rainfall. Moisture travels through the atmosphere in aerial rivers. Deforestation cuts the supply of evapotranspired water to the aerial river, disrupting the amount of rainfall downwind. Cut the trees in Ucayali, Peru, and rainfall reduces not just locally but also farther down the aerial river, in Brazil and Colombia. Stated another way, the trees curate the rain.

Ucayali is an ecologically sensitive area—an area of importance for climate change both regionally and globally. Land-use change here matters. Looking at the way water vapor moves in the air helps determine conservation hot spots. Amazon forests regulate not just the regional cycle but also the global climate system. When that land is deforested for pastureland, or to grow rice, cassava, maize, or soybeans, rainfall changes both regionally and globally.

Yet deforestation is showing no signs of slowing. In Ucayali, in the last twenty years alone, 4.7 percent of the tree cover has been destroyed.[3] Slash-and-burn agriculture and mining are largely to blame, and these extractive industries accrue profits elsewhere, yielding little for the Shipibo but small payments for unprocessed timber. The Peruvian national government is investing in roads that make logging even easier, which is likely to increase both deforestation and migration to the region. Farmacia's challenge is to keep logging, roads, mining, and other destructive forces at bay.

ECOTOURISM

The environmental and cultural goals of Farmacia are in tension with another goal, one that is more conventionally economic: to provide a source of income for the community through ecotourism. It's sort of a catch-22: in order for Farmacia to remain uncut, it has to generate income for the community by inviting people in.

Results of similar efforts in other parts of the world have been mixed. Costa Rica, after a boom of ecotourism from the 1970s to 1990s, now copes with businesses that are not environmentally sustainable taking cover in the country's green image. A similar struggle plagues New Zealand. The presence of ecotourism doesn't automatically make an entire region more ecologically sound; in fact, it can be an excuse to continue destructive practices.

The tourists that Paoyhan hopes to attract are *dieteros* interested in "dieting the plants" (going on a shamanic plant diet), and researchers who want to learn more about the biodiversity of the region.

The Farmacia committee, just a few years into their project, faces questions of great magnitude. Who will this library be for, and how can the forest generate income while still benefiting the community first? The committee is discussing the idea of allowing pharmaceutical companies to access their catalog of medicinal plants.

"The vision of the pharmacy is for our foreign friends to support us in making pills from medicinal plants. This is what we want, to support not only my fellow countrymen," Elías said. "There are a lot of foreign friends that are ill, too. We want to support them as well."

There is not yet consensus among the members of the Farmacia committee about whether attracting Big Pharma will best serve their interests. They haven't yet defined whether international pharmaceutical companies would be involved, or if the committee would be working with the industry. It is also unclear, in this scenario, whether profit would flow to Paoyhan or be diverted elsewhere. Right now, one Alianza Arkana volunteer told me, the idea of pill production is just "floating in the air."

CULTURAL SURVIVAL

Each January, many people travel from Paoyhan to Pucallpa for the Mundial, a pan-Indigenous soccer tournament. In the rainy season, and with intensified floods, it's hard to travel to many communities along the shores of the Ucayali River. Shipibo communities gather in the city of Pucallpa instead.

The Mundial is an expression of cultural survival. Spectators packed the wooden bleachers so densely that, even in the midday heat, I couldn't find a place to sit. I sought shelter from the sun in a patch of shade next to a woman with a tray full of watermelon and peanuts for sale and watched the men's soccer team from Paoyhan get knocked out by another Indigenous community in the first round. The crowd, composed of representatives from different villages around the region, was riveted by every play.

Paoyhan's women's team, playing barefoot on the dirt pitch, won and advanced. The field had a deep puddle in the center, left over from the previous night's rains, that the players navigated with skill, sometimes splashing with the ball through ankle-deep water. Alianza Arkana had screen-printed matching jerseys for the Paoyhan teams, with complex geometric designs, or *kené*, on the back, a nod to Shipibo design and the cosmovision of the plant world.

A few blocks away, at Casa Ametra, members of the Farmacia committee gathered at Alianza Arkana's kitchen between rounds of the tournament for a visioning meal, served around a long table. On the menu were boiled plantains, chicken, and rice. The stove was lit, the pots bubbling. We pulled in extra chairs and passed around salt and soda.

Nora Ractuqui Sastro, age fifty, a member of the committee, took a break from cooking to extol the virtues of Farmacia. She emphasized cures for diseases that commonly afflict foreigners. "There are many medicinal plants for all kinds of diseases," she said. "Cancer and other things. And you can go there and diet. Anything that has to do with medicinal plants, you can learn."

When she visits the library, "I feel good, happy, because the foreigners come to see and know the forests," Nora said, "and then some funds come in. That is why I am happy when people from other countries come to get to know the big trees."

When she gets older, Nora hopes that her children will replace her with their knowledge of the plants. Her favorite medicinal plant is a tree, *capirona* (*Calycophyllum spruceanum*), which she uses to heal her baby when he has diarrhea or fever. To use *capirona*, Nora cuts the bark and squeezes it. She gives her baby the liquid that is left to drink, uncooked.

The educational arm of the Living Library is just getting started. In September 2018, Alianza Arkana led a trip for Shipibo youth in Paoyhan to visit Farmacia. Fifty participants traveled in four boats, guided by the local committee. In the future, the Farmacia committee plans to make these visits more frequent, to collaborate with the school in town such that plant life is part of the curriculum.

"Many young people see plant medicines as something 'old people do,'" an Alianza Arkana blog post reported.[4] "However, after seeing certain plants and trees live for the first time, one of the young girls, Kelly Cassandra, said, 'Now I feel like I know what my mom was talking about.' If young people never experience a plant, how are they going to connect to it? How can you value what you don't know?"

16

DENMARK

RENEWABLE

Samsø is a Danish island in the Kattegat strait, an hour's ferry ride from the mainland. At forty-four square miles, with a population just over thirty-seven hundred, Samsø is known for agricultural products—potatoes and beets, mostly—and for renewable energy.[1] Since 2007, the island has been energy-positive, producing more energy from wind and biomass than the population consumes.

Until the late 1990s, Samsø relied on imported oil and coal-fueled electricity. Then, in 1997, the Danish Ministry of Environment and Energy appointed Samsø to be Denmark's renewable energy island.[2] The ministry was looking for a showcase community to prove that Denmark's Kyoto target—to cut greenhouse gas emissions by 21 percent—was possible. Søren Hermansen, a vegetable farmer from the island, was tasked with carrying out a ten-year plan for the transition to renewables.

At first, farmers were hesitant. "It was a top-down project when it started, and nobody really knew what it was all about," Søren told me. He convened meetings for the community where an engineer would describe how straw from the fields could feed the houses with heat. They looked at spreadsheets. People groaned at the math, but there was one prospective they definitely found attractive: saving money on their heating bills. It wasn't just about altruism. Over time, Søren brought up bigger questions: How do we improve local jobs? How do we save on imported fuels and insulate the community from fluctuating fuel prices? How can we use our resources for heating?

Once people saw that renewable energy would benefit their wallets, their jobs, or their businesses, more people came on board. Saying that "we're doing this to improve the climate," Søren added, was "too abstract." Climate-first messaging would have brought in "the 10 percent, hippies that were already convinced," he said, "and all the farmers would stay home." The challenge was "to find the 80 percent in the middle who didn't say much but were waiting for the right cue to come in and say yes."

The island achieved carbon neutrality in 2007, with eleven onshore and ten offshore wind turbines. Solar panels and biomass energy are also in use; many islanders have replaced their oil burners, added insulation at home, and support the island's district heating plant. The year 2007 also marked the opening of the Samsø Energy Academy, a meeting place and educational center focused on renewables. Thousands of people from around the world visit the Energy Academy each year, looking to take the lessons from Samsø back home to their communities. In 2017, having just attended a conference about the UN Sustainable Development Goals in Denmark, I was one of those visitors.

THE ENERGY ACADEMY

Søren told me that he sees part of his job as "keeping this island alive." The responsibility is broader than just his family—it's about making Samsø a place where people want to live, stay, and come home to. The population skews older. Many young people leave the island when they are fifteen or sixteen to pursue their education, and they don't come back.

"In a city," he said, "everything is in a tube. I think maybe one of the main reasons that we have a climate problem is because we have disconnected from nature. Instead of being in nature, we have separated ourselves." In an urban area, "nature is a place we go on Sundays." Islands, in contrast, generate a feeling of community—they are well defined, interdependent. On an island, "you are equally guilty or responsible for what is happening," Søren explained. This forces people to look each other in the eyes and say, how can we make it better?

The energy island project has generated some optimism. In 2017, professional opera singers from the Royal Opera in Copenhagen came to Samsø to create a project called Opera in Unlikely Places. They put on a show at the Energy Academy, outfitted with pulse meters and CO_2 meters, to see how much energy is necessary to produce an opera. A full house came to see it.

"This is the unusual you can expect on an island like this," Søren added. On an island, people are free to experiment, try things out, and figure out how to do them better. "If you did it in the Royal Opera, it would be by the book. Here, you can make mistakes and still be good, and people can laugh and have fun and get something they absolutely didn't expect."

Samsø exports their surplus energy via an underground cable that travels eleven miles to the mainland.[3] Søren laments that the market is flooded. "When it's windy, it's windy everywhere," he said. "We are looking at how we can use more energy here so we can change the pattern and have more electric cars and charge the batteries when there's abundant amounts of energy." There are rare days when Samsø imports energy: cloudy, quiet days in the wintertime when the solar panels produce nothing. "We are part of the market," Søren noted. Electrons can be traded, along with potatoes and other goods.

Søren has since turned his focus to teaching. A recent project, Samsø 100, aims to connect Samsø to one hundred other places in the world that have "the same ambition, the same pattern, the same urge to make change and create a sustainable society." The goal is to inspire people to start where they are. Communities in Australia, the Caribbean, Hawaii, and Japan are all on board. "We are part of a bigger structure," Søren told me.

LESS

Malene Annikki Lundén is the other half of the charismatic couple; she works alongside Søren at the Energy Academy to build capacity and communication. Malene has lived in Samsø for over thirty years and takes pride in the community's "courage to keep going." Meeting that first ten-year target wasn't enough; people want to press sustainability initiatives further, to experiment with a biocircular economy. "It's not a big-scale movement, but it's small steps," she said. This gives Malene a reason to smile. "It's a good, grounding feeling to live in a society where we don't make revolutions, but we take small-scale actions."

Malene hopes that young people learn to use their hands more often. "Could it be possible to be more simple?" she asked. "Less is more." Her kids moved to the city and don't plan to come back. "That's also what I work for. How can we make it more attractive to be in the countryside?" In Malene's view, when she sees young families who need to live in the city, earn a lot of money, have a car, and pay high rent, "if you look in their eyes, you can see they are pretty empty."

She wondered out loud: "What is a good life? What creates a good life?" Part of the attraction to Samsø, for her, is that there aren't a lot of cars or shops. There's not a market "pulsing in your face all the time, making you feel that you need to be a consumer." Malene owns shares in an onshore wind turbine. She is not just a consumer. "That was a big mind shift," she said. Now her task is to convince other people, in other places, that they can shift, too.

BIOGAS

Samsø's transportation sector still uses gasoline, diesel, and gas. Their next goal is to become free of fossil fuels by 2030. The biggest culprit is *Prinsesse Isabella*, the ferry to Jutland; it uses liquid natural gas imported from the Netherlands by a truck. Each year the ferry requires thirty gigawatt hours of energy. In order to become self-sufficient, Samsø needs a biogas plant—a facility that could reuse and recycle nutrients already available on the island. Smaller users could benefit, too. A pickling factory uses liquid petroleum gas for heat and could change to biogas. Farmers could also transition to using fertilizer produced at the facility.

Knud Tybirk works for the Samsø municipality, trying to implement this plan. His job, in addition to securing permissions, is to find $10 million euros in investments for the project. He described his work to me in terms of a metaphor. "You could say that the biogas plant is like one big cow. You have to feed it every day and it produces gas like a cow does but it doesn't produce milk. It produces gas and fertilizer," Knud told me in his office. He picked up a salt and a pepper shaker on the table. One cow—the salt—is fed with sludge from the pickling industry, waste potatoes, sewage sludge, and sorted organic waste from households and shops. That process produces gas and a slurry called "digestate," which can be used to fertilize grass. Then the second cow—the pepper—is fed slurry from cattle and pig production on the island, along with straw and organic waste from agriculture. That digestate becomes fertilizer for farms on the island. This creates what Knud calls a "double loop of nutrients." The biogas, which contains methane, can be used to power the ferry.

The field-to-ferry concept hasn't been tested, but liquefied natural gas is easier to transport on ships and trucks than natural gas in other forms. "Some of these restrictions that an island has create creativity, you could say, because you have to find other solutions," he said. Knud estimates that the biogas plant will create five to ten associated businesses, or at least improve business for existing companies, because their waste will now be an asset. The plant will be located opposite the pickling factory.

He noted that there are some nuisances associated with biogas—traffic, smell, and the visual impact on the landscape—that people in Samsø have voiced concerns about. "We are trying to be as open as possible and have our ears stiff and listen to what people say," Knud said.

Students on the island developed a concept for a circular-economy visitor's center at the biogas plant. The first level would be a showroom that shows how the plant integrates the carbon cycle, the nutrient cycle, and the water cycle on the island, tracking the life of a red beet or a head of red cabbage. The top floor would be a dome of glass, to grow tomatoes or cucumbers using nutrients and heat and surplus carbon dioxide from the biogas plant. The goal is "to cultivate something that people can taste, so to speak," Knud said. He dreams that perhaps tourists could climb up or rappel down the tanks.

"We could try to hide it away, but we could also actually promote it and say this is something we're proud of, that we want to show. And people drive by every day and say, okay, this is my biogas plant. I produce the energy for the ferry. I produce fertilizers for the farmers. I create jobs here. I create a sustainable island."

FIELD TO FERRY

Gunnar Mikkelsen, a project manager at Biosamfund Samsø, drove me in his electric car to see some of the agriculture-based input for the biogas digester that he is testing with local farmers.

"All waste at Samsø we are going to make into a resource," he told me, his hands on the wheel. Currently, waste from the island has to be sailed away or flows out to sea. "We're going to turn the circle, so that nutrients and water are going back to the agricultural sector." The idea is to maintain carbon sources that normally had been thrown away, and move them, through the biogas plant, back into the soil.

"Samsø farmers are like all farmers. They have to earn money. So we also are going to work out how much money you can earn when you produce energy crops," Gunnar said, pulling over beside a field. The goal is to make an energy crop as lucrative as other options, like winter rye. Though few farms are organic, Samsø's vegetables are coveted in the rest of Denmark, and considered a quality brand. This earns farmers more.

One technique is to use a "catch crop"—a fast-growing crop that steps in between plantings of the staple crop. Clover grass, for instance, takes nitrogen from the air. When mixed in with vegetables and beans, it can be cut and, with compost, used as a fertilizer.

In Samsø, Gunnar told me, winter rye is cut in May. After that, he advises the farmers to sow a mixture of maize, barley, and broad beans, with clover grass at the bottom of the field. Broad beans fix additional nitrogen from the air, which provides a surplus of nitrogen to the biogas plant. The biogas slurry that comes back to the farmer will have a higher content of nitrogen, making the next crop more prolific. The maize and rye are harvested as normal crops, but they benefit from the cooperation of the other plants and see increased yields of up to 30 percent. The biogas plant will mostly consume waste, with a bit of extra crops to supplement.

"This is only the little research plot," Gunnar told me, picking a broad bean for me to see. "Just to show: Is it possible?" The grass rustled around our ankles.

THE LITTLE MERMAID

I met Lupita Pocket at a food truck in the Copenhagen neighborhood of Nørrebro, where she was serving tacos. She grew up in Mexico City, where her family would buy big jugs of water for drinking. "We are not allowed to drink the water from the taps in Mexico," she said. Her mom would tell her, as a child, to "hurry and drink a lot of water before it gets bad."

This comment didn't make sense to Lupita, until once, after not drinking the water for a few days, "small green plants" started to grow inside the bottle. Lupita wouldn't let her mom throw it away. She thought of it as her aquarium. A few days later, "little bugs" began to squiggle around. After a week, she told me, there were fish. Lupita had her aquarium. She loved it.

When she was five years old, she fell in love with the Hans Christian Andersen story "The Little Mermaid." At the end of the book was a picture of the statue of the Little Mermaid in Copenhagen's port. Lupita decided then and there that she wanted to become a mermaid. She knew that there must be mermaids in Denmark. She promised herself that one day she would become a mermaid, too. She decided that Danish girls must go to school in the morning and in the evening, learning how to be mermaids. She wanted that life.

"When I was old enough, crazy enough, I just grabbed my things, moved here, and found out," Lupita said. One day, by chance, she met the director of an art school. "He thought I had some talent, and I was funny," Lupita said, "so he gave me a scholarship for physical theater school."

She trained as an actress, and now she performs as a mermaid at an outdoor theater in Copenhagen. The show is about redefining notions of success. "Of course, I'm not really a mermaid inside of the sea. But you know, you can form that on the street and still feel that you hit one million dollars every day," she said. "I don't care a shit about money," she added. "It's just that, hey, I did what I wanted."

17

SWEDEN

MELT

After my trip in Denmark, I meandered north to Sweden and took an eighteen-hour train from Stockholm to Abisko. On this stretch, I shared a compartment with two painters and a hiker. We started talking because, well, what else was there to do? The train rumbled past bogs and hillsides draped in blue. Then, we crossed the Arctic Circle. The train conductor announced this moment with nonchalance. Outside the window, the seasons accelerated. Snow dotted the tops of ridges nearby. I put on an extra layer. The colors in the Arctic landscape were deeper than I could have ever imagined.

Maria Olofson-Thörnqvist, one of my compartment-mates, confided that it was her first time taking a train this far north. She was on her way to an art course. Maria told me that she loves to swim. At home in Gothenburg, she lives beside Lake Siljan, one of the largest lakes in Sweden. That spring, Maria told me, the water level was lower than usual. Plants choked out her favorite swimming spots. "We were waiting for the snow to melt," she said. But even in late spring, the water level stayed low—there wasn't enough snow to make a difference. "It looked very special, like somebody had pulled out the plug," she added.

The snow nearby has also become more unpredictable. Each year, tens of thousands of people gather in March for a cross-country skiing competition. In the last few years, there hasn't been enough snow, and the event organizers have had to expend energy producing snow with machines. Some people have suggested moving the competition to February. But even the winter temperatures have been unpredictable, fluctuating between extreme cold and unseasonable warmth and rain. "It went up and down, up and down," Maria told me. "Is it just weather? Or is it a part of the climate change?" She couldn't say for sure.

BEHAVIOR

The train rumbled on, stopping in Kiruna before continuing north. Charlotte, another woman in my compartment, wanted to talk about small things. "You have to live as you want other people to do," she told me. "You can just think: What is the best for your children? Do you want to change to green energy? Do so. It's rather easy. You can do small things, and you can be happy." Considering waste and travel are other options. "You can take your bicycle everywhere and you can walk and you can take the train," she added. "You can sort your garbage and you can buy things that are produced nearby." The point is to start somewhere, and to start small.

PERMAFROST

The village of Abisko houses a scientific research station, operational since 1913.[1] It was there I met Joachim Jansen, then a PhD student at Stockholm University focusing on how the warming climate is changing methane and carbon dioxide emissions from Arctic lakes. He took me out to his field site, Stordalen mire, a wetland in a discontinuous permafrost zone, where some areas were thawing. As the permafrost starts to thaw, cracks open and new pools form or merge into the pools nearby. The link between thawing permafrost and greenhouse gas emissions from lakes is not well understood. "As the water starts to warm, emission of greenhouse gases from the lake increases, creating a positive feedback to climate warming. In areas with thawing permafrost, organic carbon from the soil can enter the lake and contribute to further emissions," Joachim explained. "We are interested in how fast that goes, so we try to put a number on it. If it warms one degree, how much extra methane is going to come out of the landscape and come out of the lakes?"

To measure that, Joachim paddled a rowboat. On the lakes and in the sediment, bacteria or archaea produce methane gas in bubbles. The bubbles accumulate and then rise. "On a hot day, you can actually see the bubbling," he told me. Joachim took funnels from a hardware store, turned them upside down, and created a trap for the gas to enter. He set up forty traps on the lakes.

The complexity of the climate system is what keeps Joachim going. "The main thing was figuring out how it works," he told me. "It's a very urgent problem," and one that is not fully understood. "We don't know how fast this wetland is going to emit methane when we warm it one degree," Joachim said. "That's terrifying." His goal is to contribute to understanding the Arctic feedback system so that future predictions can be more accurate.

Joachim encourages people from other latitudes to make the journey north. "I think it's important for people that live in places where they are not affected by climate change to come and visit those places that are, because it is visible already here," he said. In Abisko, the permafrost is thawing. "Places where you could walk last year, you can't walk this year. And for us that's not a big issue, but for the people that herd reindeer, the Indigenous people of the region, that means that they have to move their stock somewhere else." Frequent snowmelts in winter create hard layers of ice where the reindeer used to be able to find food. "People that live in places that are affected by climate change have a valuable perspective, and you don't get that perspective if you don't go and visit those places," he said.

We walked across the bog, following a trail outlined with planks of wood, hauled a rowboat onto one of the lakes, and paddled out to the sample sites, conducting the day-to-day work of climate science. Joachim took diligent notes in his notebook. On the way back, we gathered wild blueberries to eat. That night, we sat beside Lake Torneträsk to watch the aurora borealis in greens and purples—a curtain rippling overhead.

ADAPTABILITY

For Keith Larson, a scientist at Umeå University, sustainability is a myth—but adaptability is what humans do best. I accompanied Keith on his fieldwork at

Abisko National Park. On the slopes of Nuolja, a hundred-year-old research program has been following plant life relative to snowmelt dates. The project, started by Swedish botanist Theodor Fries, allows scientists today to compare tree-line data going back decades. In the last century, trees have ascended 130 meters up the slope, moving in tandem with the degree and a half of warming due to climate change. Boreal plant species have moved into this region that didn't exist, or were less common, a hundred years ago.

Keith hammered stakes into the ground, replacing the wooden poles that had disintegrated over time, and fortifying the research for future generations. "We anticipate that climate change is going to be an issue for at least another hundred years," Keith told me, and that this study would be repeated again after he was gone. "These are the kinds of things you can't answer in two to three years," he added.

Sustainability, in Keith's mind, is a human invention, a fantasy. "In some ways it's a corporate myth, because sustainability is just not something that human beings do," he said. "Our ability to adapt to climate change is constrained by the fact that there are people living almost everywhere now. So that's a problem. But it's a problem that can be overcome." The key to adaptability, Keith told me, is to have people pay the true cost of things—a way of incentivizing renewable energy and traveling low and slow, instead of by plane. "I think we need to focus on how we move on from the paradigm of growth," he said. Keith swung the mallet over his head, preparing for the next hundred years of research.

MONITORING

Living in parallel with machines are the people who fix them. I met Robert Holden, a research engineer and polar technician, in Abisko. Robert works for the Swedish Polar Research Secretariat on a project that monitors greenhouse gases all over Europe. In addition to making repairs, he goes on monitoring expeditions to Greenland and the North Pole. In Abisko, Robert measures greenhouse gas fluctuations coming out of the discontinuous permafrost peatland.

Farther north, the permafrost is unbroken. Farther south, there isn't any. Small changes in temperature have a large impact in this borderland.

During his time at the station, Robert witnessed Arctic permafrost melting and collapsing, indicating a larger trend of global climate change. There are dry areas in the mire, a specific kind of wetland, where the water in the soil is frozen and lifts up the surface, creating something solid to walk on. When the winters are even slightly warmer, the permafrost collapses into a wet hole. Eventually, permafrost is projected to disappear from Abisko entirely. This thaw releases greenhouse gases like carbon dioxide and methane into the atmosphere, compounding the problem by supplying more planet-heating emissions.

"If you were to try to tell something to people who doubt climate science, what would you say?" I asked.

"The thing we're doing wrong is trying to explain the evidence to people who aren't listening," Robert said. "I have colleagues here who think it's immoral to have children because they're going to be a burden to the world. It's these things we never explain. We always explain the science with no emotion, but it's actually just really painful." Here Robert took a beat to breathe. "It almost feels a bit pointless."

"What keeps you going?" I asked.

"I'd like us to change."

Robert described his love of ice and glaciers, with a focus on color and texture. "You can sit and stare at it forever," he said. "As it's freezing you get these crazy patterns and a million colors of blue and gray. It's very beautiful, very peaceful." The colors and shapes of the ice provide him with a deep sense of calm. It is this he fears losing, along with everything else.

18

NORWAY

PLASTIC

Already north of the Arctic Circle, I caught a bus to Tromsø, Norway, following the places where scientists gather. Tromsø is often the last stop that researchers make before crossing the Arctic Ocean to Svalbard, the northernmost year-round settlement in the world, and home to researchers of many nationalities. Geir Wing Gabrielsen, a senior research scientist in environmental pollutants at the Norwegian Polar Institute in Tromsø, sat down with me in his office for an interview. Stacks of papers and books crowded most of the available desk space, and a wide window overlooked the water. Geir has been researching Arctic animals for nearly four decades. In recent years, his focus has turned to plastic pollution.

In 1987, Geir started investigating the diet of the fulmar, a bird that can live up to sixty years in the wild. Of the forty birds he sliced open, four had plastic in their stomach. (Fulmars are surface feeders, so they likely think the plastic is plankton). In 2013, he repeated the study, with dramatically different results. Some birds had more than two hundred pieces of plastic in their stomachs, preventing the uptake of nutrients. In Europe, fulmars have been found on the beaches, starved to death because of the overload of plastic in their stomachs.

Part of the reason there's so much plastic in the Arctic is that ocean currents are changing. Geir showed me a graph of carbon dioxide measurements taken at the Mauna Loa Observatory in Hawaii, one of the longest CO_2 time series in the world, starting in 1958.[1] While there is an annual cyclicality between a Northern Hemisphere summer and winter (lower in the summer, when carbon dioxide is absorbed by the sea), the measurements show that the amount of CO_2 has been steadily increasing since the time series began, surpassing four hundred parts per million in 2013. And an increased concentration of greenhouse gases in the atmosphere—carbon dioxide, nitrogen oxide, aerosols, and methane—is making the oceans warmer, which changes their currents. This, in turn, pushes more plastic contamination into the Arctic. Plastic production is, itself, a fossil-fuel-intensive process. Two kilograms of oil are used to produce one kilogram of plastic.[2]

Plastic is now found not only in Arctic surface waters, but also on the bottom of the ocean floor, and in sea ice. Geir, in his work with seabirds, has witnessed other changes in the ecosystem. Fjords that used to be dominated by polar species now have Atlantic species. Species that used to be farther south, like capelin, herring, mackerel, and Atlantic cod, are more prominent than polar cod. The ambient temperature of the fjord in the summer has increased by 3–4 degrees Celsius. In the winter, sea water is 1–2 degrees higher than the old average. These physical changes eventually translate into biological changes.

Atlantic water contains higher levels of contaminants than Arctic water. Svalbard used to be surrounded by polar water; now Atlantic water dominates, which means more persistent organic pollutants, plus mercury. These substances are transported by the current, but also by the animals themselves. They are stored in body fat, which impacts their immune and endocrine systems.

When the Atlantic system drifts northward, pollution more readily enters the food chain. Fish eat the plankton, the seal eats the fish, the polar bear eats the seal, and the toxicity accumulates in the body of the apex predator at the top of the food chain. Changes in atmospheric and sea currents are accelerating the transportation of contaminants throughout the system.

Geir pointed to the Great Pacific Garbage Patch, a swirling mass of plastic more than three times the size of France, located between North America and Asia.[3] "It's telling us we need to do something to prevent what's going on with regard to plastic pollution," he said. There are an estimated one hundred thousand marine mammals and turtles and one million seabirds killed by plastic pollution each year.[4] "We all agree to take care of our coastline, but nobody wants to take care of what's going on far away from us, out at sea," Geir said.

SVALBARD

Alena Dekhtyareva grew up in Murmansk, an industrial city in northwestern Russia, where the sky was lined with smokestacks. This inspired her to study environmental engineering, and work to help reduce the city's environmental impact. When we met at the University of Tromsø, Alena was a PhD student in environmental surveillance technology, focusing on air pollution in the Arctic. "One might think that the Arctic is a clean place. There is no pollution. There are almost no people there. But it's not true," she told me. There's a mixture of long-range pollution from the midlatitudes and local pollution in Svalbard, an archipelago in the Arctic Ocean with four permanently occupied scientific research stations.

Alena described the archipelago as a small community of around two thousand researchers. Sometimes, the lack of trees felt depressing. "It's either snow or it's bare soil," she said. But soon, she acclimated to the landscape. "For me, it feels almost like a second home after so many visits."

Svalbard houses two coal power plants, which provide the research stations with heat and energy. Alena monitored pollution with equipment that takes samples of air and analyzes the content inside for concentrations of different compounds. She also registered particles and gases emitted from cruise ships that carry tourists. These greenhouse gases trap warm air, amplifying climate change.

PEOPLE VS. OIL

When we spoke in 2017, Martine was fifteen years old. She grew up in Tromsø and was part of a group of youth with the environmental organizations Natur og Ungdom and Greenpeace suing the Norwegian government for opening up new areas for oil and gas drilling in the Arctic. They viewed this as a violation of Article 112 in the Norwegian constitution, which states:

> Every person has the right to an environment that is conducive to health and to a natural environment whose productivity and diversity are maintained. Natural resources shall be managed on the basis of comprehensive long-term considerations which will safeguard this right for future generations as well.
>
> In order to safeguard their right in accordance with the foregoing paragraph, citizens are entitled to information on the state of the natural environment and on the effects of any encroachment on nature that is planned or carried out.
>
> The authorities of the state shall take measures for the implementation of these principles.[5]

The court ruled in favor of the government in 2018.[6] The climate activists appealed the ruling but were unsuccessful.[7] Norway is repeatedly ranked among the wealthiest countries in the world, and the majority of that wealth comes from oil and gas.

Martine told me that some of Norway's environmental inaction could change with more youth representation in government. "I want youth in our government to have a say in decisions that are going to be a part of our future. When decisions are made, they're made by adults who are not going to live with the future that they're creating," she said. Reaching the goals of the Paris Agreement "isn't going to happen if we drill for more oil. We need to stop drilling oil and focus on the green shift and thinking more environmentally."

At that time, Lofoten, a biodiverse archipelago in northwestern Norway known for its cold-water reefs, pods of whales, colonies of seabirds, and spawning grounds for cod, was a potential ground for oil drilling. By 2019, the push for drilling in Lofoten had stopped.[8] But when we spoke in 2017, it was still an open question. Martine was concerned. "I was there this summer. I couldn't even imagine an oil platform in the middle of the sunset. That would destroy the moment," she told me. "It's absurd to even consider petroleum activity in this area."

OIL

Katrine Boel Gregussen is a representative in her late twenties for Norway's Socialist Left Party. "I'm really afraid of how the future's going to look if we're digging for oil," she told me at her kitchen table. "It's a matter of money. We have been living good because of the oil in Norway for so many years, but now we know so much better." The only way to continue to drill for fossil fuels, she added, is to blatantly disregard the future impacts. She described her country as "addicted"—not only to oil, also to the wealth that it brings.

"I think one of the biggest problems here in Norway is that we still have people who don't believe in environmental changes," Katrine added. "It's a big problem because if people don't believe in something that is so well documented, how can you get them to believe in anything? I think you don't get a good discussion if people don't believe in the facts." She finds these conversations frustrating.

In 2016, Katrine traveled to Greenland, a country where people see climate-related changes daily. "The ice is melting, and there's less fish and a lot of weather problems," she said. "It was so good to be in a country where everyone believes in the facts." For Katrine, the wealth that Norway has accumulated from oil can make it easy for people to turn away from the negative impacts of climate change. "We are one of the richest countries in the world, and I really don't think that we do enough for the rest of the world. In Norway, I think that it's easy to close our eyes," she told me. "We're so lucky to live in such a safe place, and I don't think we do enough for those who aren't as lucky as we are."

She is heartened by the generation of environmentalists coming up behind her. "They understand that this is no joke," she said. "I think that does a lot for the generations that come after us." Katrine maintains hope that even though oil is so inextricably tied to Norwegian identity, "we'll move in the right direction."

19

TURKEY

WATER CANNON

Sometimes, face-to-face conversations spin in the most unexpected directions. In Istanbul, just a few hours into wandering with a new translation of the cardboard sign, I met a gamer and graphic designer in his early twenties named Efe. Efe, a former foreign exchange student in the United States, spoke in rapid-fire English, urgency behind every word. He told me that before 2013, he had never attended a protest. That year he heard about a sit-in at Gezi Park next to Taksim Square, a protest against the redevelopment of one of the last green spaces in central Istanbul.[1] The aim of the redevelopment project was to ease congestion around the square and build a shopping center and a mosque over part of the park.

Taksim means "divide" in Turkish; the square used to be the site of Istanbul's main reservoir, where water was divided.[2] One night, police tear-gassed the protesters as they slept. This turned the small environmental protest into a swelling antigovernment movement. Long-simmering frustration against Prime Minister Recep Tayyip Erdoğan and the Justice and Development Party spilled over.

Though Efe was twenty-four and "not that politicized," he decided to join the protest. "I couldn't hold myself inside, so I just pushed to push myself into it," Efe told me. He found himself shoulder to shoulder with other people, singing and chanting in the street. Police used water cannons to disperse the crowd. The pressure was so strong that one direct hit could push a person twenty meters.

When he was young, Efe was an avid swimmer. Later, he worked as a lifeguard. But for him, this was easily the biggest water story of his life. "The irony is you cannot live without water, right? But they are using it as a weapon against you."

He told me that it felt like "a really powerful wet punch that pushes you really, really far away and you cannot resist." In the water's force, he felt like he was drowning. The bruise on his chest stayed black for two weeks. Other people had broken ribs and arms. "Thank god it was summer," Efe said. He dried off quickly.

At first, it was scary. The vehicles that shot the water were huge, intimidating. Then, at some point, Efe's fear transformed into something else. He started to joke with the people around him. "It's hot. Let's go against some water cannons to cool off," he said. The violence became something mundane, surreal: just getting wet. He learned to approach the vehicle, rather than run from it, because at the right angle, the hoses couldn't reach him.

"We got a bit experienced," he said. "That was maybe the best time of my life." He has the date of the protests tattooed on his arm.

GUEST

Ismail Tezcan walked up to me at the Kadıköy Ferry Terminal. Salt permeated the air. He wanted to speak about hospitality. In the past, Ismail said, people used to travel mostly by foot. One day, a boy arrived and sat outside Ismail's front door.

"Of course, since he is such a traveler, we asked: Are you hungry? Are you in need of a shelter? Are you thirsty? Are you sleepless?"

The boy said that he slept by the water. Ismail understood this to mean that he was hungry. "We gave him food. That's it. That is our story," Ismail said. He boarded a boat and left across the Bosphorus, the ferry churning bubbles in its wake.

RAMADAN

While wandering through Istanbul, I ducked into a Starbucks. Hülya Eren stood in front of me in line. Seeing my sign, she insisted on buying me an espresso.

We sat across from each other at a table outside, sipping. "I don't have a big story about water," Hülya said, readjusting the pins on her hijab. "We are Muslim. We fast for thirty days as Allah ordered us. During these thirty days, you cannot eat or drink water at certain times."

Abstaining from water reminds Hülya of how precious and valuable it is. "In the month of Ramadan, we understand the blessing of Allah much better," she explained. "Water is even more important than food." When the weather is hot, the thirst becomes insufferable. "Water is precious," she said.

BALANCE

The intersection of Üzerlik Sokağı and Muvakkithane Caddesi in Istanbul is a pedestrian thoroughfare. People sell sneakers and wristwatches and handbags on the street. Şerif Canyurt, a man from Malatya with a prodigious beard, hawked sunglasses. He leaned against the awning of a doorway.

"The balance of the world has changed," he told me. "Past times were better. When I was a kid, I grew up in streams. I grew up in fresh water. It was more natural then."

Now, he lamented, "there is construction everywhere, everywhere. The forests are in a bad situation." His hometown of Malatya struggles with water pollution from nearby factories.

Sometimes the fish die en masse; no one knows why.

"Mankind is very ungrateful," he concluded.

KIZ KULESI

Long ago, as legend would have it, two people were in love. One was a Byzantine princess, and her father didn't like the match; the man she fell for wasn't royal enough. In order to prevent the lovers from seeing each other, he shut his daughter in a tower on a small island at the entrance to the Bosphorus: Kız Kulesi. She was alone, confined by water, but found a way to communicate with light.

"Every night she lit a fire to guide her lover's way, and he swam to her," Deniz Tekant, then a high school student in Istanbul, told me. One night, a horrible storm came—a storm so strong that it burned out the fire.

The lover lost his way and drowned. His remains washed onto the shores of Kız Kulesi.

"When the princess saw him, she decided that she couldn't live without him," Deniz said. The princess committed suicide, her only means of escape.

DEVELOPMENT

When she was a child in Istanbul, Nermin Şahin fetched water for her family from a stream. "Our feet were always muddy," she told me. "So we tried to find a solution." Her family paid to have water delivered, which they then transferred into a tank at home. But that solution, too, was impermanent. Finally, they drilled a well thirty or forty meters deep. "That water is still there now," she told me. "It's ice-cold but it is not drinkable."

The well didn't solve the problem, but Istanbul's water infrastructure slowly developed. "Water was connected to our more beautiful homes. And let's say we started living like human beings."

HOPE

When Yaman Tekant was a child in Izmir in the 1960s and '70s, his family weathered shortages of both water and electricity. "This was not unique to Izmir," he told me. "All parts of Turkey were experiencing similar problems."

In order to cope with the shortage, his family stored water in lidded buckets so that even if the water wasn't running in the morning, they would have some to bathe in. Electricity shutoffs would be announced in advance, but the water was more erratic. It could be hours or days before it returned, "so we had to plan for it," Yaman told me. "This went on for years."

Yaman developed a practice of conserving water that stays with him to this day. He turns on the water for only the minimum amount of time necessary when brushing his teeth, shaving, or taking a bath. In years where there was less rain, the problem was worse. The municipality tried to jump-start rainfall with cloud seeding. "But it was mostly unsuccessful," he said, laughing, "or just a very insignificant amount of rain would fall, but that would really not help."

Since 2000, rain has been more frequent, which cut down on Turkey's water shortages. Sometimes, in the short term, he told me, a climate impact can be positive.

"I'm aware of the warnings for the future," Yaman said. "I'm sure that many mistakes are being made and that politicians are not planning far ahead." He worries that water shortages will return.

"I really can't see a very bright future," he said. "Especially my generation, perhaps despite certain problems during our childhood, maybe we lived through the best years of the Earth. I'm worried about my children, even my children who are already grown up, for their future," he said.

"If I have any grandchildren, I'm very worried," he added. "What will their hope be for their future?"

20

UNITED KINGDOM

POND

Bodies of water can be good to think with. Since 2014, artist Clare Whistler has been convening weeklong events in different parts of Pevensey Levels in Sussex, where people reflect on different aspects of water. One of the attendees, Pete, comes every year to talk about ponds and biodiversity. He emphasizes that "a pond can be the size of a bucket," Clare told me, "and I've been having this go around my head for quite a long time."

Then, in July 2016, Clare woke up feeling quite low about the environment. She decided that the only cure for her malaise was to make a pond. She phoned up Pete, who dutifully came over to her garden. Pete chose a spot where a few trees had been taken down because they had died of disease. He planned the pond so that the plants and bugs would be native. They designed an irregular shape, dug a hole, and put a black, woven piece of plastic lining on the ground, followed by another layer of earth. Clare took a garden hose and water from her system and filled it to the top. "It's a very quick process," she said, "Things go back, or go forward, very quickly when they're left alone." The rewilding had begun.

Three weeks later, Clare was sitting by the pond when the first dragonfly arrived. "If that isn't joy or hope, I don't know what is," she said. "Make a hole in the ground and it explodes. Life explodes."

That summer, there was a drought. Clare watched the water level go down, fretting. Pete told her not to worry. Then in the autumn, once it had filled up, the pond was full of leaves. “Should I clear the leaves?” Clare asked Pete. He responded, “What happens to a pond in a forest?” Slowly, she stopped worrying. She relinquished control.

Then, Clare began to think about thinking. “You begin thinking about ponds and pondering and how to ponder and why still water is different from rushing water and why there might be times in life where a pond and still water is what may be demanded,” she noted.

The next gathering she held would be Pond Week, a series of meetings at different ponds where people could have pond conversations, “which really means just sitting around very low to the ground and looking into the pond, and seeing what arises.”

After a few practice ponderings, one of Clare’s friends decided to make a pond in her own garden. Another attendee was less enthused—it turns out his brother had drowned in a pond before he was born. “You don’t know what conversation will come up,” Clare told me. Each subject is a starting place.

“For me, the arts—the creation of beauty—is probably the best way to deal with this changing environment.”

BUSINESS

Karen Brennan is the group CFO for Futerra Sustainability Communications, a creative agency with headquarters in London that aims to make sustainable development so desirable that it becomes normal. We first connected at COP22 in Morocco and stayed in touch. Karen firmly believes that businesses have a role in bringing climate solutions to the table.

“I am a chartered accountant with a degree in finance and accounting, and I can assure you that in all the studies I’ve ever done, and some of them twenty, thirty years ago even, no one ever talked about businesses as just being for profit,” she said. “It was always around considering where you were geographically and economically and the people that worked for you, as well as your shareholders. So I’m not quite sure when the agenda was hijacked to just be about shareholder value and the one piece of information on the bottom line." Businesses, she added, "are operating in their communities for a reason. There's more than just profit at stake."

Karen noted that the companies she has worked with have figured out that they need a purpose in order to bring their customers with them. "When you can align your purpose with the community in which you operate, it starts to make sense and it starts to become natural and easy," she told me.

"And it shouldn't be a fight. It shouldn't be an opposing force," Karen added. "The role of business, in my view, is to live in harmony and balance within the societies that we have created. Every business needs to take that seriously."

COMMUNITY

A mile from the city center, tucked on council land as part of a green corridor in Leeds, the Bedford Fields Community Forest Garden is home to hundreds of species of plants, many of which are edible. The forest garden is run by local volunteers. Ben Lawson joined in 2013 and took on responsibility for developing and managing the garden. He estimates that there are six community forest gardens in the UK, which he defines as "this weird hybrid of being an educational space, a community space, and being a place where people can forage and learn about forest gardening. They're totally experimental."

Ben lives fifty yards from the garden. "It's my main place of nutrition," he told me. He makes a fifty-plant pesto that he sells at the local market to raise funds for the garden. Surviving on plants from the garden "gets quite complex. It destroys the idea of a normal salad," Ben said. "Salads aren't the same once you start forest gardening. They're much more interesting."

Forest gardening as a movement started about thirty years ago in Britain, but it has deeper roots in countries closer to the equator. In the UK, the main concern is a lack of light—plants have to be spaced farther apart in order to prosper.

The garden is, in its own way, a response to climate change—a way of making the community more resilient. "Forest gardening, for what you put in and what you get out, it's very efficient. You put in very little land and you get quite a lot out. If you go look at industrial pig farming, for example, you'll see that they throw everything at it and get, you know, proportionally not as much out," Ben told me.

Forest gardening is a low-maintenance system. He uses perennial plants and never turns the soil. Mulching and using ground covers mean that there's very little weeding necessary, too.

Forest gardens can take up to twenty years to mature in productivity. Ben described the Leeds garden as a "petulant child, at the moment." Rhubarb, carrots, raspberries, strawberries, apples, pears, and sour cherries all do well. Birds thrive among the vegetation. And, as an added benefit, the plants don't require any water aside from the rain. After the initial period of design, and with some minimal maintenance, "it just looks after itself."

WATER BIRTH

I met the Wilsons on their sheep farm on the Knoydart Peninsula in far-north Scotland. No roads connect the nearest community, Inverie, to the mainland. The only way in is by boat, on a seven-mile sea crossing, or on foot, hiking 18 miles. I opted for the boat route.

Anna Wilson and I became friends while I was cycling in New Zealand in 2015. Anna contacted me about a year later to say that she was returning home to Scotland to have a baby. Three weeks after she gave birth to Ossian, I happened to be in the UK. I took a five-hour train from Glasgow to Mallaig, and then a ferry across the water to Knoydart, a community of a hundred. Anna asked me to bring groceries: broccoli, carrots, cauliflower, bananas, and goat milk.

She had planned to deliver at the hospital, but a week before her due date, her water broke in the middle of the night. Anna was at home on her family's farm. That night, it was raining and hailing. The sea was stormy, "which meant that we weren't going to be going anywhere," she told me. She labored for a few hours, wandering around the house. After six hours, she felt like she needed to push. Anna sent her father to get her friend Heather from the village at Inverie. "I just hoped that they were going to get back in time."

Her mother ran a bath with candles and essential oils. "It wasn't long before he was nearly here, but I managed to kind of slow things down until Dad and Heather got back from the village," Anna said.

Fifteen minutes later, Ossian's head popped out, followed by his shoulders. "I reached down and looked at him through the water and lifted him up and out of the water and into my arms. And after that was my fourth water, which was tears, which we all had of joy that he'd arrived safe and sound and there were no complications."

TRICKSTER

In Inverie, there's the wharf, one tearoom and pottery shop, one school building, a row of homes facing the water, what might be the world's most remote pub, and a single road connecting it all.

Anna's parents live a half-hour bumpy drive from Inverie on a dirt road that snakes through steep hillsides and dramatic winter yellows and blues. One night Anna's father, Iain, took me in his truck to see the Northern Lights. It was raining, and the clouds cleared just for a moment to reveal a smudge of the galloping green. You get the sense, being here, that the people are shaped by the landscape. Remoteness fosters self-sufficiency. When it rains, water collects, running downhill, pooling in the lowest places. Iain loaned me a pair of gumboots to keep my feet dry.

The story he wanted to share was a legend from the nearby hills. "A part of the farm is a mountain called Beinn Na Caillich. *Beinn Na Caillich* means, in English, 'the Hill of the Old Woman,'" he told me. The old woman in question lived on the border between two estates, at a fork in the mountain stream. Rivers are a dividing line. Arbitrary, yes, but also possible to manipulate.

When the tax collector came, the woman would go up to the stream above her house, put a few stones in, and divert the water so that she lived on the other estate. When the tax collector arrived from the other estate, she went back up to the stream, took a few stones out, and diverted the water down the other side of her house, implying that she lived in the other estate.

"They tried to arrive unannounced, but news always traveled fast in the highlands," Iain told me. "So she never paid taxes all her days, or rent."

PHOSPHORESCENCE

Iain's wife, Jo Wilson, lives at the edge of the sea. "The weather plays such a big part in our lives, and you see it. You see it coming and you see it going," she told me.

One day, about a decade before we spoke, Jo's neighbors knocked on the door late at night. "You must go down to the water's edge and see the phosphorescence," they said. "It's just incredible."

Jo and Iain walked to the water's edge. They splashed in the sparkling plankton.

"Every time you moved the water, it was like having sparkles and glowing throughout. It was just incredible," she told me. "We splashed around and ran about in the water like little kids. All the trails of the sparkling phosphorescence followed us in every movement that we made."

CONCLUSION

STORIES ARE DOORS

This will be, like any story, incomplete. If you don't see yourself or your experience with water and climate change reflected in these pages, I would love to hear it. You can tell your own story—or record someone else's—at 1001stories.org.

I believe that we make the world through our actions. In the face of a challenge as large and universal as climate change, we need all kinds of people to listen and join. Yes, it can feel overwhelming. But as long as these stories prompt some kind of shift—no matter how small—then we are moving forward. And movement is what I'm all about.

What can you do? Start where you are. Start small. To quote an organizing toolkit from the People's Climate Movement: "To change everything, it takes everyone."[1]

REPETITION

In college I took a course on race, gender, and performance that introduced me to the work of Anna Deavere Smith, an actress, playwright, and professor who continues to inspire me. Back in the 1980s, she began interviewing and recording stories from people across the United States. Then she translated those interviews, verbatim, into performances. In her 2005 TED Talk, she outlined her creative process.

When she was young, her grandfather told her, "If you say a word often enough, it becomes you."[2] In interviewing people, Anna hoped to walk in their words—to absorb what she called the "organic poems" of America.

We tell ourselves stories in order to survive—to make sense of the chaos of the present moment, and of all the moments that came before. Storytelling is a kind of world-building. If we take the time to deeply understand the world we have, including all its flaws and imperfections, we can more effectively create the future that we want.

SLOW CYCLING

The bicycle is a tool for connection. A conversation starter. A tool for transcending borders, whether of nationality or gender or class or age. In listening, I give the whole of myself—my ears, my heart—to a storyteller. In cycling, I give the whole of myself—my body, my spirit—to a place. I move through the landscape, and the landscape moves through me.

Slowness has become part of my daily practice. Slow cycling means waving at people as I pass. Slow cycling means traveling with a Sharpie in my handlebar bag and writing messages on telephone poles or highway guardrails: "Just play." / "Slow is beautiful." / "Read more poetry." Slow cycling means, on some days, stopping every mile to add another line to the poem that I am churning in my head. Slow cycling means picking a flower at the summit of a mountain pass, putting it behind my ear, and getting off the bicycle to do a happy dance because I made it to the top of a mountain using the power of my own body.

For me, cycling is a form of active listening. Even when I had to trade in the bicycle for other forms of travel, I tried to integrate that same sense of wonder, of unpredictability, into my journey. Every day was a new start. There is always space for joy. There is always space for questions.

Climate change and water are two issues that are so beautifully entangled in everything. The biggest problems are never easy to parse. By slowing down enough to listen, we can begin to unravel the regional and textural complexities in a way that might point us toward solutions.

HOW TO LISTEN

Listening again to years' worth of material, I can hear myself becoming a better interviewer and asking better questions over time. This was, indirectly, my path into journalism—I realized that reporters and editors get to ask and listen all the time, and then communicate what they learned to the world. That act of translation was exciting for me: more accessible than poetry, more impactful (I hope) than academia.

But to keep this tradition of oral history strong, we need to remind people that their voices matter. In order to share a story, a person has to trust you first. You have to make them feel comfortable. There are many ways of doing this. Informed consent is important. Never turn on an audio recorder unless someone says it is okay. People need to know where the material is going, and why you're doing it. Take down people's contact information so that you can get in touch with them later.

Every story is a gift, and you should treat it as such. After someone tells you their story, make a point of thanking them. Invariably, after the audio recorder is turned off, people will ask: "Is that okay? Was my story all right?" I say yes. I say thank you. Let people know that their voice matters, because it always does.

When we stop to listen to one another, sometimes magic happens. Empathy develops. We create bonds across lines of difference. Listening is an act of love. And if there is one thing that we need more of in the dialogue about climate change, it is compassion. Take a moment to fall forward into the rhythm of someone else's story. You never know where it might lead.

ACKNOWLEDGMENTS

Writing a book is a team sport. I'm immensely grateful to all the storytellers who have propelled me around this planet multiple times. I couldn't have kept going without the kindness and generosity of the 1,001 strangers who took the time to share a piece of their lives with me, offered me a spare racecar bed to spend the night, or a nourishing meal, many glasses of water, a tent spot outside a monastery, or a couch on which to rest. Thank you for your stories and incredible hospitality. This book would not exist without you. I hope that I can pay it forward.

I am indebted to the many people who, at an early stage of this project, believed in its value. Thank you to the Office for the Arts at Harvard for the Artist Development Fellowship, to the Gardner & Shaw Postgraduate Traveling Fellowship from Harvard, the Council for the Arts at MIT, and the National Geographic Society for supporting segments of this journey.

I once opened a box of raisins that read on the inside flap: "Be grateful for good teachers all your life." Thanks are due to Deborah Foster, Maria Tatar, Josh Bell, Jorie Graham, Kimberley Patton, Ajantha Subramanian, Andrew Littlejohn, Sally Morris, Becky Moore, Emily Gray, Caroline Guerra-Baker, Robin Bernstein, Helen Mirra, Ernst Karel, and every teacher whom I have been fortunate enough to learn from.

A big thanks to Samia Bouzid, Anna Woorim Chung, and Jeff DelViscio for their work on the 1,001 Stories map. Thank you to Warich Ngamkanjanarat for translations from Thai and Julaluck Intasin for translations from Lao.

Erica Plouffe Lazure, Brittany Montgomery, Geri Diorio, Nana Chen, Kaitlyn Gerber, Miranda Shugars, Anna Rasshivkina, Julie Zauzmer, Leo Schwartz, and Caroline Catlin all read early drafts of this book—thank you. Thanks to Eric Cervini, for the encouragement via text, and to the many Binders, for teaching me how to pitch. To the typo brigade—Julia Carmel, Ellen Shakespear, Karin Stevens, Elena Gosalvez Blanco, Nancy Coleman, Viveca Morris, Leah Varjacques, Amanda Chisholm, Noah Goodwin, David Wolinsky, Anna Merrens, Melis Tekant, Caroline Lowe, Tamar Nisbett, Alexandra E. Petri, Denise Colaianni, Ralph Baskin, Brandy Machado, Brie Belz, Cassandra Euphrat Weston, Lou Berl, Chris Berl, Eliza MacLean, Valerie Piro, Nicole Virgo-Carter, Kip Clark, Sasha Baskin, Anne Martin, Dayna Reggero, Christina Sullivan, Eve Driver, Bailey Richardson, Brian Mukhaya, Cornelia Waterfall—you all are fabulous.

Immense gratitude to Sam Ford at Tiller Press, for taking a chance on me, and to Emily Carleton and Samantha Lubash, editors extraordinaire. Patrick Sullivan designed the gorgeous cover. Many thanks to my fantastic agent, Tim Wojcik, for believing in this book, even when we received what felt like 1,001 first-round rejections with the market wisdom that books about climate change don't sell. Thanks to Greta Thunberg for changing the conversation—she opened the door wide enough for the rest of us to step through.

Thank you to *The New York Times* Opinion section, especially to Jim Dao, Clay Risen, Jenée Desmond-Harris, Nana Asfour, Honor Jones, and Chris Conway, all of whom shaped my editing and writing in different ways. Thanks to the *Rest of World* team, for your support and camaraderie.

Big love to my family—Emmett, Camrynn, Mom, Dad, Charlie—for giving me the space to write, encouragement to adventure, and a warm nest to return to. And thanks to my grandfather, Stu Howkins, for teaching me how to ride a bike. We practiced balance in the high school parking lot and then you let go. I travel chasing that feeling of freedom in motion.

Thank you, dear reader, for making it to the end.

NOTES

Introduction

1. Conor Friedersdorf, "The Mistake Authorities Made After the Boston Marathon Bombing," April 24, 2015, https://www.theatlantic.com/politics/archive/2015/04/shutting-down-boston-was-the-wrong-decision/391388/.
2. In Washington, DC, I went back to using balloons. I recorded stories inside of a balloon booth, a blue box made of long balloons, six feet wide and three feet deep, twisted by myself and Julie Zauzmer, a *Washington Post* reporter who works on weekends twisting balloons for birthday parties and other events. But that was years later. We made the balloon sculpture in her apartment and carried it to Malcolm X Park.
3. Rejection is fuel. Every application I fill out makes me better at completing the next one. The process of writing about what I'm doing helps me understand it. Writing is thinking, refining, whittling. Someone told me that it takes a hundred tries for every success (sometimes less). But the one-hundred-tries mentality has helped me to not be afraid of failing.

4. Margaret Wheatley, "Listening as Healing," *Shambhala Sun*, December 2001, https://www.margaretwheatley.com/articles/listeninghealing.html.
5. "Drinking-Water," World Health Organization, July 14, 2019, https://www.who.int/news-room/fact-sheets/detail/drinking-water.
6. "Human Right to Water and Sanitation," UN Water, United Nations Department of Economic and Social Affairs, April 29, 2014. https://www.un.org/waterforlifedecade/human_right_to_water.shtml.
7. "Historic New Sustainable Development Agenda Unanimously Adopted by 193 UN Members," Sustainable Development Goals, United Nations, September 25, 2015. https://www.un.org/sustainabledevelopment/blog/2015/09/historic-new-sustainable-development-agenda-unanimously-adopted-by-193-un-members/.
8. "Mortality and Burden of Disease from Water and Sanitation," Global Health Observatory (GHO) data, World Health Organization, August 29, 2018. https://www.who.int/gho/phe/water_sanitation/burden_text/en/.
9. "Goal 6 | Ensure Availability and Sustainable Management of Water and Sanitation for All," Department of Economic and Social Affairs, United Nations, n.d., https://sdgs.un.org/goals/goal6.

Chapter 1: Tuvalu

1. Marcus Woo, "King Tides: What Explains High Water Threatening Global Coasts?" *National Geographic*, January 30, 2014, https://www.nationalgeographic.com/news/2014/1/140130-king-tides-high-tides-sea-level-rise-science/.
2. "Current and Future Climate of Tuvalu," International Climate Change Adaptation Initiative, Pacific Climate Change Science Program, 2011, https://world.350.org/pacific/files/2014/01/4_PCCSP_Tuvalu_8pp.pdf.
3. Alan Taylor, "A Visit to Tuvalu, Surrounded by the Rising Pacific," *The Atlantic*, August 15, 2018, https://www.theatlantic.com/photo/2018/08/a-visit-to-tuvalu-surrounded-by-the-rising-pacific/567622/.
4. "Current and Future Climate of Tuvalu," International Climate Change

Adaptation Initiative, Pacific Climate Change Science Program, 2011, https://world.350.org/pacific/files/2014/01/4_PCCSP_Tuvalu_8pp.pdf.

5. "Information about: Pacific Access Category Resident Visa," Immigration New Zealand, accessed January 21, 2021, https://www.immigration.govt.nz/new-zealand-visas/apply-for-a-visa/visa-factsheet/pacific-access-category-resident-visa.
6. Lianne Dalziel, "Government Announces Pacific Access Scheme," Beehive.govt.nz, December 20, 2001, https://www.beehive.govt.nz/release/government-announces-pacific-access-scheme.
7. "CO_2 Emissions (Metric Tons per Capita)," The World Bank Data, Carbon Dioxide Information Analysis Center, accessed January 21, 2021, https://data.worldbank.org/indicator/EN.ATM.CO2E.PC.

Chapter 2: Fiji

1. "Millennium Development Goals (MDGs)," World Health Organization, February 19, 2018. https://www.who.int/news-room/fact-sheets/detail/millennium-development-goals-(mdgs).
2. "Fleet Growing for 'Mua'—the Sailing Voyage to World Parks Congress 2014," IUCN, July 17, 2014, https://www.iucn.org/content/fleet-growing-mua-sailing-voyage-world-parks-congress-2014%20.
3. "Mua Voyage," YouTube, Ocean+ TV, November 18, 2014, https://www.youtube.com/watch?v=bTShZrbsBKM&ab_channel=Ocean%2BTV+.

Chapter 3: New Zealand

1. "Where Your Water Comes From," Watercare, accessed January 21, 2021, https://www.watercare.co.nz/Water-and-wastewater/Where-your-water-comes-from.
2. "Auckland City (Census 96) (1996 Census of Population and Dwellings)," Statistics New Zealand, accessed January 21, 2021, https://cdm20045.contentdm.oclc.org/digital/collection/p20045coll18/id/72.

3. "2018 Census Population and Dwelling Counts: Stats NZ," 2018 Census population and dwelling counts | Stats NZ, September 22, 2019, https://www.stats.govt.nz/information-releases/2018-census-population-and-dwelling-counts.
4. Auckland water efficiency strategy, Watercare, 2017, https://www.watercare.co.nz/CMSPages/GetAzureFile.aspx?path=~%5Cwatercarepublicweb%5Cmedia%5Cwatercare-media-library%5Creports-and-publications%5Cwater_efficiency_-strategy.pdf&hash=cc41b37c9c7aa6b9ad0c9148c640f9a9776819c5d8b990928b478ecac7e410e8.
5. David Fickling, "Farmers Raise Stink over New Zealand 'Fart Tax,'" *Guardian*, September 4, 2003, https://www.theguardian.com/world/2003/sep/05/australia.davidfickling.
6. David Fickling, "Plan to Halt Wind in the Woolies," *Guardian*, June 22, 2004, https://www.theguardian.com/world/2004/jun/23/australia.davidfickling.
7. Kerryn Pollock, "Waikawau Tunnel and Beach," *Te Ara Encyclopedia of New Zealand*, Ministry for Culture and Heritage Te Manatu Taonga, March 30, 2015, https://teara.govt.nz/en/photograph/34643/waikawau-tunnel-and-beach.
8. "ECO Regional Gathering—Taranaki's Beauty and the Beast," Climate Justice Taranaki, January 30, 2015, https://climatejusticetaranaki.files.wordpress.com/2014/11/taranaki-eco-regional-gathering-for-all-v3.pdf.
9. "New Zealand's Oil & Gas History," Energy Mix, Petroleum Exploration & Production Association of New Zealand, accessed January 22, 2021, https://www.energymix.co.nz/our-resource/new-zealands-oil-and-gas-history/.
10. Roger Gregg and Carl Walrond, "Alpha Well, Moturoa," *Te Ara Encyclopedia of New Zealand*, Ministry for Culture and Heritage Te Manatu Taonga, June 12, 2006, https://teara.govt.nz/en/photograph/8919/alpha-well-moturoa.
11. "Evaluating the Environmental Impacts of Fracking in New Zealand: An Interim Report," Parliamentary Commissioner for the Environment,

November 2012, https://www.pce.parliament.nz/media/1241/fracking-interim-web.pdf.

12. Anne MacLennan, "Fracking, Climate and Health," Climate Justice Taranaki, Ora Taiao—NZ Climate & Health Council, January 15, 2015, https://climatejusticetaranaki.files.wordpress.com/2012/11/fracking-health-and-climate-anne-maclennan-feb15.pdf.
13. Emma Hatton, "Toxic Algal Blooms Still Flourishing despite the Cold," RNZ, August 12, 2018, https://www.rnz.co.nz/news/national/363921/toxic-algal-blooms-still-flourishing-despite-the-cold.
14. "New Zealand's Greenhouse Gas Inventory 1990–2018," Ministry for the Environment, April 2020, https://www.mfe.govt.nz/sites/default/files/media/Climate%20Change/new-zealands-greenhouse-gas-inventory-1990-2018-snapshot.pdf.
15. "Population," Stats NZ, accessed January 22, 2021, https://www.stats.govt.nz/topics/population.
16. Mike Joy and Sylvie McLean, "Despite Its Green Image, NZ Has World's Highest Proportion of Species at Risk," *The Conversation*, April 30, 2020, https://theconversation.com/despite-its-green-image-nz-has-worlds-highest-proportion-of-species-at-risk-116063.
17. "Status of Our Freshwater Species," Ministry for the Environment, 2017, https://www.mfe.govt.nz/publications/fresh-water/fresh-water-report-2017-ecosystems-habitats-and-species/status-of-our%20.
18. Diana Beaglehole, "In and around Whanganui," *Te Ara Encyclopedia of New Zealand*, Ministry for Culture and Heritage Te Manatu Taonga, June 16, 2008, https://teara.govt.nz/en/whanganui-places/page-2.
19. Tanea Tangaroa, "Submission to Environmental Protection Agency—Hazardous Substances," Environmental Protection Agency, 2016, https://www.epa.govt.nz/assets/Uploads/Documents/Hazardous-Substances/HS-WSR-Completed-Consultation/Submissions/HPC-2016/aaa7a111b2/Submission-No.57-HPC-2016-Tanea-Tangaroa.pdf.
20. "Resource Management Act 1991," Resource Management Act 1991 No 69 (as of September 30, 2020), Public Act Contents, September 30,

2020, https://www.legislation.govt.nz/act/public/1991/0069/latest/DLM230265.html.

21. Tui Gilling, "Whanganui Waterways Scoping Report—A Report for the Waitangi Tribunal," Ministry of Justice, November 2001, https://forms.justice.govt.nz/search/Documents/WT/wt_DOC_94177816/Wai%20903%2C%20A029.pdf.
22. Laurel Stowell, "Transforming a Whanganui Wasteland to a Wetland," *Whanganui Chronicle* (*NZ Herald*, April 8, 2017), https://www.nzherald.co.nz/the-country/news/transforming-a-whanganui-wasteland-to-a-wetland/QRGMSY2B5SEWQUR3D5WGHYHCYM/.
23. "Ice Sheets," National Science Foundation, accessed January 24, 2021, https://www.nsf.gov/geo/opp/antarct/science/icesheet.jsp.
24. "Understanding Sea Level," Sea Level Change Portal, NASA, n.d., https://sealevel.nasa.gov/understanding-sea-level/global-sea-level/ice-melt.
25. Robert Kunzig, "Climate Milestone: Earth's CO2 Level Passes 400 Ppm," *National Geographic News*, May 12, 2013, https://www.nationalgeographic.com/news/energy/2013/05/130510-earth-co2-milestone-400-ppm/.
26. "End of Summer Snowline Survey," NIWA, 2018, https://niwa.co.nz/climate/research-projects/climate-present-and-past/southern-alps-glaciers/end-of-summer-snowline-survey.
27. "Tasman Glacier Retreats," The Earth Observatory, NASA, March 24, 2017, https://earthobservatory.nasa.gov/images/89901/tasman-glacier-retreats.
28. Mike Williams, "Icebergs off NZ Probably Came from the Other Side of Antarctica," NIWA, November 7, 2006, https://niwa.co.nz/news/icebergs-nz-probably-came-other-side-antarctica.
29. Michael Field, "Iceberg Continues Journey Past New Zealand," *Stuff*, December 7, 2011, https://www.stuff.co.nz/world/6101620/Iceberg-continues-journey-past-New-Zealand.
30. "Sheep Shearing on an Iceberg," Natural History New Zealand Moving Images, April 3, 2008, http://www.nhnzmovingimages.com/archives/page/3/article/105/.
31. Simon Hartley, "Solid Energy Mothballs Briquette Plant," *Otago Daily*

Times Online News, October 15, 2013, https://www.odt.co.nz/business/solid-energy-mothballs-briquette-plant.

32. Jane Young, "A Voice For Nature," South Otago Forest & Bird, January 2012, https://www.yumpu.com/en/document/read/20855623/a-voice-for-nature-jan-12pdf-forest-and-bird.
33. Phil McCarthy, "Mataura Briquette Plant Could Be Sold, Dismantled," *Stuff*, March 4, 2015, https://www.stuff.co.nz/business/industries/66933264/mataura-briquette-plant-could-be-sold-dismantled.

Chapter 4: Australia

1. "Bushfire—Black Saturday, Victoria, 2009," Australian Disaster Resilience Knowledge Hub, Australian Institute for Disaster Resilience, 2019, https://knowledge.aidr.org.au/resources/bushfire-black-saturday-victoria-2009/.
2. Dale Cox, "Supermankind," Dale Cox Art, March 2020, https://dalecox.com.au/collections/supermankind/.
3. "Millennium Drought Report," Environment, Land, Water and Planning, August 26, 2019, https://www.water.vic.gov.au/climate-change/millennium-drought-report.
4. triple j, "Camp Flozza," October 11, 2011, https://www.youtube.com/watch?v=8SsQrXhNwdI&ab_channel=triplej.
5. "Timeline: The Rise and Fall of Gunns," ABC News, September 24, 2012, https://www.abc.net.au/news/2012-09-25/gunns-timber-company-rise-fall-timeline/4235708.
6. "Forestry Tasmania—Utter Spite at Camp Flozza," The Habitat Advocate (Deep Silent blog), October 8, 2011, https://www.habitatadvocate.com.au/forestry-spite-at-camp-flozza/.
7. Barrie May et al., "Tasmanian Forest Carbon Study Full Report," CO_2 Australia Limited, July 31, 2012, http://www.dpac.tas.gov.au/__data/assets/pdf_file/0007/172789/ForestCarbonStudy_Report.pdf.
8. Miranda Gibson, "I Spent 449 Days in a Tree without Touching the Ground—It Was All Worth It," *Guardian*, June 25, 2013, https://www

.theguardian.com/commentisfree/2013/jun/25/tasmania-tree-protest-logging.

9. Andrew Darby, "Tasmania's Old Growth Forests Win Heritage Protection," *Sydney Morning Herald*, June 24, 2013, https://www.smh.com.au/national/tasmanias-old-growth-forests-win-heritage-protection-20130624-2os3p.html.
10. "Leadbeater's Possum," Environment, Land, Water and Planning, September 7, 2020, https://www.wildlife.vic.gov.au/our-wildlife/leadbeaters-possum.
11. "Zooxanthellae and Coral Bleaching," *Smithsonian*, accessed January 22, 2021, https://ocean.si.edu/ocean-life/invertebrates/zooxanthellae-and-coral-bleaching.
12. Katie L. Cramer et al., "Widespread Loss of Caribbean Acroporid Corals Was Underway before Coral Bleaching and Disease Outbreaks," *Science Advances*, April 22, 2020, https://advances.sciencemag.org/content/6/17/eaax9395.
13. "Coral Bleaching Events," Australian Institute of Marine Science, accessed January 22, 2021, https://www.aims.gov.au/docs/research/climate-change/coral-bleaching/bleaching-events.html.
14. "Species and Climate Change: More than just the Polar Bear," The IUCN Red List of Threatened Species, Species Survival Commission, 2009, https://www.iucn.org/downloads/species_and_climate_change_1.pdf.
15. "The Value of the Reef," Great Barrier Reef Foundation, accessed January 22, 2021, https://www.barrierreef.org/the-reef/the-value.
16. "Plants," Department of Agriculture, Water and Environment, Australian Antarctic Division, February 7, 2018, http://www.antarctica.gov.au/media/news/2011?a=7893.
17. Sharon A. Robinson et al., "Rapid Change in East Antarctic Terrestrial Vegetation in Response to Regional Drying," *Nature Climate Change*, September 24, 2018, https://www.nature.com/articles/s41558-018-0280-0.
18. "Gallery: Brisbane's 1974 Floods, 40 Years On," ABC News, January 26, 2014,

https://www.abc.net.au/news/2014-01-27/brisbane-1974-floods-40-year-anniversary/5197952.

19. "Climate Change Pushing Australia's Platypus towards Extinction: Researchers," Phys.org, January 20, 2020, https://phys.org/news/2020-01-climate-australia-platypus-extinction.html.
20. Latika Bourke, "LNP Backbencher George Christensen Likens Climate Change to Science Fiction Film Plot," ABC News, July 8, 2014, https://www.abc.net.au/news/2014-07-09/backbencher-likens-climate-change-to-science-fiction-film-plot/5583734?nw=0.
21. Latika Bourke, "Nationals MP George Christensen Calls Green Activists 'Terrorists,'" *Sydney Morning Herald*, September 25, 2014, https://www.smh.com.au/politics/federal/nationals-mp-george-christensen-calls-green-activists-terrorists-20140925-10lt5a.html.
22. "About Us," NRG Gladstone Operating Services, accessed January 22, 2021, https://www.nrggos.com.au/.
23. "Renewables," Department of Industry, Science, Energy and Resources, accessed January 22, 2021, https://www.energy.gov.au/data/renewables.
24. "Basics," Geoscience Australia, accessed January 22, 2021, https://www.ga.gov.au/scientific-topics/energy/basics.
25. John W. Miller, "The $200,000-a-Year Mine Worker," *Wall Street Journal*, November 16, 2011, https://www.wsj.com/articles/SB10001424052970204621904577016172350869312.
26. "Port of Hay Point," North Queensland Bulk Ports Corporation Ltd, accessed January 22, 2021, https://nqbp.com.au/our-ports/hay-point.
27. "Rangers Confirm Sighting of Second Croc in Mary River," *Fraser Coast Chronicle*, July 18, 2013, https://www.frasercoastchronicle.com.au/news/rangers-confirm-sighting-second-croc-mary-river/1950246/.

Chapter 5: At Sea

1. Chris Watson, *Beyond Flying: Rethinking Air Travel in a Globally Connected World* (Cambridge: Green Books, 2014).
2. "WWF Footprint Calculator," WWF, accessed January 22, 2021, https://footprint.wwf.org.uk/.
3. S.W. Boggs, "'This Hemisphere,'" *Journal of Geography*, June 23, 2008, https://www.tandfonline.com/doi/abs/10.1080/00221344508986498.
4. Brian J. Cudahy, "The Containership Revolution: Malcom McLean's 1956 Innovation Goes Global," World Shipping Council, September 2006, https://www.worldshipping.org/pdf/container_ship_revolution.pdf.
5. "History of Containerization," World Shipping Council, accessed January 22, 2021, https://www.worldshipping.org/about-the-industry/history-of-containerization.
6. Elizabeth Knight, "International Air Fares at 30-Year Low," *Sydney Morning Herald*, April 22, 2016, https://www.smh.com.au/business/companies/international-air-fares-at-30year-low-20160422-gocr1r.html.

Chapter 6: Thailand

1. Everton Fox, "Thailand Hit by Its Worst Drought in Decades," Climate News, *Al Jazeera*, March 30, 2016, https://www.aljazeera.com/news/2016/3/30/thailand-hit-by-its-worst-drought-in-decades.
2. Elizabeth Preston, "These Shrimp Leave the Safety of Water and Walk on Land. But Why?," *New York Times*, November 18, 2020, https://www.nytimes.com/2020/11/18/science/shrimp-parade-thailand.html.
3. Sahana Ghosh, "What Is Cloud Seeding?," Mongabay, August 1, 2019, https://india.mongabay.com/2019/08/what-is-cloud-seeding/.
4. Amanda Little, "Weather on Demand: Making It Rain Is Now a Global Business," Bloomberg, October 28, 2015, https://www.bloomberg.com/features/2015-cloud-seeding-india/.

Chapter 7: Laos

1. "Vat Phou and Associated Ancient Settlements within the Champasak Cultural Landscape," UNESCO World Heritage Centre, accessed January 22, 2021, https://whc.unesco.org/en/list/481/.

Chapter 8: Cambodia

1. Rebecca Solnit, *Men Explain Things to Me* (Haymarket Books, 2014).
2. Mong Palatino, "Animals Can't Escape Cambodia's Worst Drought in 50 Years, Either," Global Voices, May 9, 2016, https://globalvoices.org/2016/05/09/animals-cant-escape-cambodias-worst-drought-in-50-years-either/.
3. Lauren Crothers, "Animals Die as Cambodia Is Gripped by Worst Drought in Decades," *Guardian*, May 4, 2016, https://www.theguardian.com/global-development/2016/may/05/animals-die-cambodia-worst-drought-decades.
4. "Community Ravaged by Drought Steps up to Save Monkeys," *Khmer Times*, May 9, 2016, https://www.khmertimeskh.com/23660/community-ravaged-by-drought-steps-up-to-save-monkeys/.

Chapter 9: China

1. Carla Freeman, "Quenching the Thirsty Dragon: The South-North Water Transfer Project–Old Plumbing for New China?," Wilson Center, accessed January 22, 2021, https://www.wilsoncenter.org/publication/quenching-the-thirsty-dragon-the-south-north-water-transfer-project-old-plumbing-for-new.
2. "Records Set by South-to-North Water Diversion Project," China Internet Information Center, People's Republic of China, May 8, 2018. http://www.china.org.cn/waterdiversion/special/2018-05/08/content_51178944.htm.

3. "China's New Grand Canal Brings Water to Arid North," *China Daily*, December 16, 2014. https://www.chinadaily.com.cn/china/2014-12/16/content_19093414.htm.

Chapter 11: Kyrgyzstan

1. Katie Arnold, "In Kyrgyzstan, Warming Brings Less Water—and More Conflict," Reuters, November 9, 2018, https://www.reuters.com/article/us-kyrgyzstan-water-climatechange/in-kyrgyzstan-warming-brings-less-water-and-more-conflict-idUSKCN1NE0BW.
2. Malika Giles, "UNESCO Recognizes Kyrgyz Epic of Manas," *Moscow Times*, December 8, 2013, https://www.themoscowtimes.com/2013/12/08/unesco-recognizes-kyrgyz-epic-of-manas-a30280.

Chapter 12: Morocco—UN Climate Talks

1. Rahma Sophia Rachdi, "The Bab Ighli Village of COP 22 Is Almost Ready to Welcome 15,000 People," United States Press Agency News, November 3, 2016, https://www.uspa24.com/bericht-9859/the-bab-ighli-village-of-cop22-is-almost-ready-to-welcome-15-000-people.html.
2. "Marrakech Conference Carries Momentum of Climate Action: News: SDG Knowledge Hub: IISD," SDG Knowledge Hub, November 21, 2016, http://sdg.iisd.org/news/marrakech-conference-carries-momentum-of-climate-action/.
3. "Needs Inventory," Center for Nonviolent Communication, 2005, https://www.cnvc.org/training/resource/needs-inventory.
4. Alister Doyle and Meredith Davis, "Trump Win Boosts Coal, Hits Renewable Stocks," Reuters, November 9, 2016, https://www.reuters.com/article/us-usa-election-climatechange-idUSKBN1342E0.
5. "Environmental Reactions to Trump Victory," *Deutsche Welle*, November 9, 2016, https://www.dw.com/en/environmental-reactions-to-trump-victory/g-36321582.

6. Anna Pérez Catalá, "Trump, El Cambio Climático y La COP22," *El País*, November 14, 2016, https://elpais.com/elpais/2016/11/10/3500_millones/1478797750_123573.html.
7. James Delingpole, "Trump Victory Totally Rains on UN Climate Summit's Parade," Breitbart, November 10, 2016, https://www.breitbart.com/europe/2016/11/10/trump-victory-totally-rains-un-climate-parade/.
8. Andy Costa, "My Dream for Africa Foundation," 2021, http://mydreamforafrica.org/fr/.
9. Mimi Kirk, "Africa's First Bike-Share Just Launched in Morocco," *CityLab*, Bloomberg, November 11, 2016, https://www.bloomberg.com/news/articles/2016-11-11/africa-s-first-bike-share-just-launched-in-morocco.
10. "Ugandan Community Leader Receives Top Forestry Prize," Food and Agriculture Organization of the United Nations, September 10, 2015, http://www.fao.org/news/story/en/item/328330/icode/.
11. Hanna Gunnarsson, "Gertrude Kabusimbi Kenyangi," Women2030, March 6, 2018, https://www.women2030.org/gertrude-kabusimbi-kenyangi/.
12. James Gorman, "Is This the World's Most Diverse National Park?," *New York Times*, May 22, 2018, https://www.nytimes.com/2018/05/22/science/bolivia-madidi-national-park.html.
13. Miriam Telma Jemio, "El Chepete y El Bala, La Amenaza a Las Comunidades Del Madidi y El Pilón Lajas," Indigenas, 2019, https://indigenas.lapublica.org.bo/pilon-lajas-madidi/el-chepete-y-el-bala/.
14. Jim Christie, "How Stockton Went Broke: A 15-Year Spending Binge," Reuters, July 3, 2012, https://www.reuters.com/article/us-stockton-bankruptcy-cause/how-stockton-went-broke-a-15-year-spending-binge-idUSBRE8621DL20120703.
15. "The Delta," California Department of Water Resources, accessed January 22, 2021, https://water.ca.gov/Water-Basics/The-Delta.

Chapter 13: United States

1. "Curatorial Overview," Brooklyn Museum, accessed January 22, 2021, https://www.brooklynmuseum.org/eascfa/dinner_party/curatorial_overview.
2. Glenn Albrecht, "Solastalgia: a New Concept in Human Health and Identity," PAN Partners, 2005, https://www.worldcat.org/title/solastalgia-a-new-concept-in-human-health-and-identity/oclc/993784860.
3. Anthony Leiserowitz et al., "Global Warming's Six Americas," Yale Program on Climate Change Communication, accessed January 22, 2021, https://climatecommunication.yale.edu/about/projects/global-warmings-six-americas/.
4. Leiserowitz, Anthony, Jennifer Marlon, Xinran Wang, Parrish Bergquist, Matthew Goldberg, John Kotcher, Edward Maibach, and Seth Rosenthal, "Global Warming's Six Americas in 2020," Yale Program on Climate Change Communication, October 11, 2020. https://climatecommunication.yale.edu/publications/global-warmings-six-americas-in-2020/.
5. Dana Nuccitelli, "California Just Had Its Worst Drought in over 1200 Years, as Temperatures and Risks Rise," *Guardian*, December 8, 2014, https://www.theguardian.com/environment/climate-consensus-97-per-cent/2014/dec/08/california-just-had-its-worst-drought-in-over-1200-years.
6. Ker Than, "Causes of California Drought Linked to Climate Change, Stanford Scientists Say," Stanford University, September 30, 2014, https://news.stanford.edu/news/2014/september/drought-climate-change-092914.html.
7. Danny Lewis, "The Paper Boats of Troy," *Hakai Magazine*, April 15, 2015, https://www.hakaimagazine.com/features/paper-boats-troy/.
8. AJ Valente, "Changes in Print Paper During the 19th Century," Paper Antiquities, 2010, https://docs.lib.purdue.edu/cgi/viewcontent.cgi?article=1124&context=charleston.
9. "Building the Fleet," SeaChange: We All Live Downstream, 2014, https://seachange2014.tumblr.com/the-fleet.

10. Paul Eckert, "Pennsylvania Hit by Huge Flooding, Towns Submerged," September 8, 2011, https://www.reuters.com/article/us-usa-flooding/pennsylvania-hit-by-huge-flooding-towns-submerged-idUSTRE7874I620110909.
11. Steve Stanne, "Perfect Storms: How Hurricane Irene and Tropical Storm Lee Slammed NY," *New York State Conservationist*, August 2012, https://wri.cals.cornell.edu/sites/wri.cals.cornell.edu/files/shared/documents/0812perfectstorms.pdf.
12. Christopher M. Reddy, "Oil in the Ocean: Oil in Our Coastal Back Yard," Woods Hole Oceanographic Institution, *Oceanus Magazine*, October 13, 2004, https://www.whoi.edu/oilinocean/page.do?pid=52295&tid=282&cid=2471&print=this.
13. Justin Gillis and Leslie Kaufman, "After Oil Spills, Hidden Damage Can Last for Years," *New York Times*, July 17, 2010, https://www.nytimes.com/2010/07/18/science/earth/18enviro.html.
14. David Rosenberg, "Walkabout Theater Presents A PERSEPHONE PAGEANT—An Outdoor Spectacle!," Chicago, IL, *Patch*, August 1, 2017, https://patch.com/illinois/chicago/walkabout-theater-presents-persephone-pageant-outdoor-spectacle.
15. "Perfluorochemicals (PFCs) Fact Sheet," CDC Environmental Health, November 2009, https://www.cdc.gov/biomonitoring/pdf/PFCs_FactSheet.pdf.
16. "Coakley Landfill/Pro-Wash Car Wash—North Hampton, Hampton, and Greenland, NH," The PFAS Project Lab, 2019, https://pfasproject.com/hampton-north-hampton-nh/.
17. "COAKLEY LANDFILL Site Profile," Environmental Protection Agency, accessed January 24, 2021, https://cumulis.epa.gov/supercpad/SiteProfiles/index.cfm?fuseaction=second.Cleanup&id=0101107.
18. Kyle Bagenstose, "Slow Crawl of Bucks, Montgomery County Water Contamination Lawsuits Continues," *Bucks County Courier Times*, March 9, 2018, https://www.buckscountycouriertimes.com/news/20180309/slow-crawl-of-bucks-montgomery-county-water-contamination-lawsuits-continues.

19. Mark Eichmann, "Delaware Urges Residents Using Private Wells to Get Water Tested," WHYY, March 27, 2018, https://whyy.org/articles/delaware-urges-residents-using-private-wells-to-get-water-tested/.
20. John Tunison, "Michigan Fire Marshal Wants Data on PFAS-Laden Firefighting Foam," Mlive.com, March 29, 2018, https://www.mlive.com/news/grand-rapids/2018/03/statewide_survey_on_pfas-laden.html.
21. "NH - HB1799," BillTrack50, December 12, 2017, https://www.billtrack50.com/BillDetail/900582.
22. "NH HB1214," BillTrack50, November 7, 2017, https://www.billtrack50.com/BillDetail/896131.
23. "NH HB1751," BillTrack50, November 20, 2017, https://www.billtrack50.com/BillDetail/897048.
24. "M 7.6 - Pakistan," USGS Earthquake Hazards Program, 2016, https://earthquake.usgs.gov/earthquakes/eventpage/usp000e12e/executive.
25. Rebecca Lindsey, "Mapping the 2005 Kashmir Earthquake," NASA, accessed January 24, 2021, https://earthobservatory.nasa.gov/images/35576/mapping-the-2005-kashmir-earthquake.

Chapter 14: Canada

1. "Census Profile, 2016 Census Igloolik, Hamlet [Census Subdivision], Nunavut and Yukon [Territory]," Statistics Canada (Government of Canada, August 9, 2019), https://www12.statcan.gc.ca/census-recensement/2016/dp-pd/prof/details/page.cfm?Lang=E&Geo1=CSD&Code1=6204012&Geo2=PR&Code2=60&Data=Count&SearchText=Nunavut&SearchType=Begins&SearchPR=01&B1=All.
2. Leanna Garfield, "Food Prices Are Insanely High in Rural Canada, Where Ketchup Costs $14 and Sunny D Costs $29," *Business Insider*, September 21, 2017, https://www.businessinsider.com/food-prices-high-northern-canada-2017-9#cost-effective-solutions-remain-elusive-building-greenhouses-is-one-option-but-this-method-can-beexpensive-and

-require-large-amounts-of-energy-especially-for-hydroponic-farms -thatrun-on-leds-9.

3. Ashifa Kassam, "'He Died a Hero': Canadian Man Mauled by Polar Bear While Protecting His Family," *Guardian*, July 5, 2018, https://www.theguardian.com/world/2018/jul/05/canada-nunavut-aaron-gibbons-mauled-polar-bear-kids.
4. Kristin H. Westdal, Jeff W. Higdon, and Steven H. Ferguson, "Attitudes of Nunavut Inuit toward Killer Whales (Orcinus Orca)," *Arctic*, September 2013, http://pubs.aina.ucalgary.ca/arctic/Arctic66-3-279.pdf.
5. "Conservation of Polar Bears in Canada," Environment and Climate Change Canada, November 19, 2020, https://www.canada.ca/en/services/environment/wildlife-plants-species/wildlife-habitat-conservation/conservation-polar-bears.html.
6. Chaim Christiana Andersen and Geoff Rayner-Canham, "The Ulu: Chemistry and Inuit Women's Culture," *Chem13 News Magazine*, March 5, 2019, https://uwaterloo.ca/chem13-news-magazine/march-2019/feature/ulu-chemistry-and-inuit-womens-culture.
7. "Key Findings," The Qikiqtani Truth Commission, accessed January 24, 2021, https://www.qtcommission.ca/en/key-findings.
8. Sheila Watt-Cloutier, *The Right to Be Cold: One Woman's Fight to Protect the Arctic and Save the Planet from Climate Change* (Minneapolis, MN: University of Minnesota Press, 2018).
9. "Population Size and Growth in Canada: Key Results from the 2016 Census," Statistics Canada, February 2, 2017, https://www150.statcan.gc.ca/n1/daily-quotidien/170208/dq170208a-eng.htm.
10. "Canada's Population Estimates Second Quarter 2020 (Preliminary)," Nunavut Bureau of Statistics, September 29, 2020, https://www.gov.nu.ca/sites/default/files/nunavut_and_canada_population_estimates_stats update_second_quarter_2020.pdf.

Chapter 15: Peru

1. Lyndsie Bourgon, "Indigenous People Battle Squatters and Timber Poachers in Peru's Amazon," *National Geographic*, April 12, 2019, https://www.nationalgeographic.com/environment/2019/04/indigenous-people-battle-squatters-timber-poachers-peruvian-amazon/.
2. Tom Phillips, "Brazilian Explorers Search 'Medicine Factory' to Save Lives and Rainforest," *Guardian*, April 27, 2009, https://www.theguardian.com/environment/2009/apr/27/amazon-rainforest-medicine.
3. "Ucayali, Peru Deforestation Rates & Statistics: GFW," Global Forest Watch, accessed January 24, 2021, https://www.globalforestwatch.org/dashboards/country/PER/26.
4. "Getting to Know Our Roots and Ourselves—Continuing to Plant Seeds of Change with the Youth of Paoyhan," Alianza Arkana, October 8, 2018, https://alianzaarkana.org/blog/en/2018/10/getting-to-know-our-roots-and-ourselves-continuing-to-plant-seeds-of-change-with-the-youth-of-paoyhan/.

Chapter 16: Denmark

1. Rick Noack, "In Denmark, the Forest Is the New Classroom," *Washington Post*, September 16, 2020, https://www.washingtonpost.com/world/2020/09/16/outdoor-school-coronavirus-denmark-europe-forest/.
2. Dyani Lewis, "Energy Positive: How Denmark's Samsø Island Switched to Zero Carbon," *Guardian*, February 23, 2017, https://www.theguardian.com/sustainable-business/2017/feb/24/energy-positive-how-denmarks-sams-island-switched-to-zero-carbon.
3. Diane Cardwell, "Green-Energy Inspiration Off the Coast of Denmark," *New York Times*, January 17, 2015, https://www.nytimes.com/2015/01/18/business/energy-environment/green-energy-inspiration-from-samso-denmark.html.

Chapter 17: Sweden

1. "Abisko Scientific Research Station," Interact International Network for Terrestrial Research and Monitoring in the Arctic, European Commission, 2017. https://eu-interact.org/field-sites/abisko-scientific-resarch-station/.

Chapter 18: Norway

1. "Trends in Atmospheric Carbon Dioxide—Monthly Average Mauna Loa CO_2," Global Monitoring Laboratory—Carbon Cycle Greenhouse Gases, January 12, 2021, https://www.esrl.noaa.gov/gmd/ccgg/trends/.
2. Benoit Cushman-Roisin, "Industrial Ecology—Some Useful Numbers", n.d., https://www.dartmouth.edu/~cushman/courses/engs171/UsefulNumbers.pdf.
3. Marian Liu, "Great Pacific Garbage Patch Now Three Times the Size of France," CNN, March 23, 2018, https://www.cnn.com/2018/03/23/world/plastic-great-pacific-garbage-patch-intl/index.html.
4. "Facts and Figures on Marine Pollution," Blueprint for The Future We Want, United Nations Educational, Scientific and Cultural Organization, 2017. http://www.unesco.org/new/en/natural-sciences/ioc-oceans/focus-areas/rio-20-ocean/blueprint-for-the-future-we-want/marine-pollution/facts-and-figures-on-marine-pollution/.
5. "International Law and Article 112 of the Norwegian Constitution on the Right to Environment," The Faculty of Law, University of Oslo, April 13, 2018, https://www.jus.uio.no/english/research/areas/companies/events/2018/workshop15-16-mai-article-112.html.
6. "The Court Ruled in Favour of the Norwegian State in the Lawsuit by Greenpeace and Natur Og Ungdom," Kvale Advokatfirma DA, February 2018, https://www.kvale.no/en/climate-greenpeace-and-natur-og-ungdom-sued-the-norwegian-state/.
7. "Verdict in the Appeal Lawsuit Concerning Section 112 of the Consti-

tution," Government.no, January 23, 2020, https://www.regjeringen.no/en/aktuelt/medhold-for-staten-i-ankesaken-om-grunnloven—112/id2687211/.

7. Mikael Holter, "Norway's Oil Industry Is Dealt a Stinging Blow," Bloomberg, April 6, 2019, https://www.bloomberg.com/news/articles/2019-04-06/big-oil-loses-norway-labor-party-ally-on-exploring-off-lofoten.

Chapter 19: Turkey

1. "Turkey Police Clash with Istanbul Gezi Park Protesters," BBC News, May 31, 2013, https://www.bbc.com/news/world-europe-22732139.
2. Susannah Cullinane, "Beyond the Riot Zone: Why Taksim Square Matters to Turks," CNN, June 7, 2013, https://www.cnn.com/2013/06/07/world/europe/turkey-taksim-square-symbol/index.html.

Conclusion

1. "To Change Everything, It Takes Everyone: An Organizing Toolkit." People's Climate Movement, n.d. https://peoplesclimate.org/ittakeseveryone/.
2. Anna Deavere Smith, "Four American Characters," TED, February 2005, https://www.ted.com/talks/anna_deavere_smith_four_american_characters.

ABOUT THE AUTHOR

Devi Lockwood has written about science, climate change, and technology for *The New York Times*, *The Guardian*, *Slate*, and *The Washington Post*, among others. She spent five years traveling in twenty countries on six continents to document 1,001 stories on water and climate change, funded in part by the Gardner & Shaw postgraduate traveling fellowships from Harvard and a National Geographic Early Career Grant. Lockwood graduated Phi Beta Kappa, summa cum laude from Harvard, where she studied folklore and mythology and earned a language citation in Arabic. In 2019, she completed an MS in science writing at MIT.

Twitter: @devi_lockwood
Website: devi-lockwood.com

CPSIA information can be obtained
at www.ICGtesting.com
Printed in the USA
BVHW061511270422
635079BV00003B/3

9 781982 146733